Lecture Notes in Control and Information Sciences

Edited by A. V. Balakrishnan and M. Thoma

Vol. 1: Modelling and Identification of Distributed Parameter Systems
Proceedings, IFIP Working Conference, Rome, Italy, June 21–26, 1976
Edited by A. Ruberti
Approx. 500 pages. 1977

Vol. 2: New Trends in Systems Analysis
International Symposium, Versailles, December 13–17, 1976
Edited by A. Bensoussan and J. L. Lions
VII, 768 pages. 1977

Vol. 3: Differential Games and Applications
Proceedings of a Workshop, Enschede, Netherlands, March 16–25, 1977
Edited by P. Hagedorn, H. W. Knobloch, and G. J. Olsder
XII, 236 pages. 1977

Vol. 4: M. A. Crane, A. J. Lemoine
An Introduction to the Regenerative Method for Simulation Analysis
VII, 111 pages, 1977

Lecture Notes in Control and Information Sciences
Edited by A.V. Balakrishnan and M. Thoma

3

Differential Games and Applications

Proceedings of a Workshop
Enschede 1977

Edited by
P. Hagedorn, H. W. Knobloch, and G. J. Olsder

Springer-Verlag
Berlin Heidelberg GmbH 1977

Editors
Prof. Dr. P. Hagedorn, Darmstadt
Prof. Dr. H. W. Knobloch, Würzburg
Prof. Dr. G. J. Olsder, Hengelo (Netherlands)

With 60 Figures

ISBN 978-3-540-08407-5 ISBN 978-3-540-37179-3 (eBook)
DOI 10.1007/978-3-540-37179-3

Originally published by Springer-Verlag Berlin Heidelberg New York in 1977.

2061/3020-543210

CONTENTS

LIST OF PARTICIPANTS

AKSELROD, B., 22 Hasne, Ramatgan	Israel
BASAR, T., P.O.Box 141, Kadiköy, Istanbul	Turkey
BAYEN, H., De-Noel-Platz 9, 5000 Köln 41	Germany
BERNHARD, P., 35 Rue Saint Honoré 35, 77 Fontainebleau	France
BLAQUIERE, A., Tour 14-24, 2 Place Jussieu, Paris	France
BOCK, H.G., Wegelerstraße 6, 5300 Bonn	Germany
BRAKEMEIER, W., Holstenhofweg 85, 2000 Hamburg	Germany
BREAKWELL, J., Stanford University, Stanford,Cal.94305	USA
BÜRGER, W., Breslauer Platz 3, 6100 Darmstadt	Germany
CASE, J., Northfield Place 305, Baltimore	USA
CHRISTOPEIT, N., Adenauerstraße 24-42, 5300 Bonn	Germany
ELLIOTT, R.J., 22 Newland Park, Hull, HU 5 2 DW	England
EVERS, J., P.O.Box 217, Enschede	Netherlands
FLEMING, W., Math.Dept.Brown University, Providence	USA
FRANKENA, J.F., Wagnerlaan 11, Enschede	Netherlands
GERARD, G., Labor Automatique Théorique, Paris VII	France

GRONSKI, I., Cleveland State Univ., Cleveland, Ohio	USA
HAGEDORN, P., Auf der Letteilung 22, 6101 Seeheim	Germany
HAJEK, O., Case Western Reserve Univ., Cleveland, Ohio	USA
HUIJSMANS, R., NSP, Haagsteeg, Wageningen,	Netherlands
KAAZ, M.A., Graf-Recke-Straße 84, 4000 Düsseldorf	Germany
KNOBLOCH, H.W., Wittelsbacherplatz 2, 8700 Würzburg	Germany
KOHLMANN, M., Wegelerstraße 6, 5300 Bonn	Germany
KRAPF, H.G., Hochschulstraße 1, 6100 Darmstadt	Germany
KWAKERNAAK, H., P.O.Box 217, Enschede	Netherlands
LEITMANN, G., Univ.of California, Berkeley, Cal.94720	USA
LEPERTIER, J.P., Ronte de Laval, 72000 Le Mans	France
LEWIN, J., P.O.Box 2250, Haifa	Israel
LONGMAN, R., Columbia University,New York	USA
MARSHALL, B., Av.J.B.Clement, 93430 Villetanense	France
MARZOLLO, A., Université Paris VII, 2 Place Jussieu	France
MAURER, H., Am Hübland, 8700 Würzburg	Germany
NOVAK, A., Karlsplatz 13, A 1040 Vienna	Austria

OLSDER, G.J., Fr.Santmanstraat 14, Hengelo	Netherlands
OTTERBEIN, St. Martinstraße 46, 6100 Darmstadt	Germany
PACHTER, M., P.O.Box 395, Pretoria 0001	South Africa
PARTHASARATHY, Th., 2605 S.Indiana, nr.1907, Chicago, Illinois	USA
PIKKEMAAT, G.F., P.O.Box 800, Groningen	Netherlands
PLASMANS, J., Drabstraat 112, B 2550 Kontich	Belgium
PLEWE, K., Weyertal 86-90, 5000 Köln	Germany
POHL, F.C., Wegelerstraße 6, 5300 Bonn	Germany
REBHUHN, D., 41 Wendell Street 23, Cambridge/Mass 02138	USA
REHN, W., Rathausstraße 12, 6500 Mainz 1	Germany
RENES, I., Anthony Fokkerweg 2, Amsterdam	Netherlands
RISITO, C., Via Bezzecca 12, Parma	Italy
ROXIN, E., Univ.of Rhode Island, Kingston, R.I. 02881	USA
SACHS, E., Schloßgartenstraße 7, 6100 Darmstadt	Germany
SCHÄFER, B.E., Heinrichstraße 54, 6108 Weiterstadt	Germany
SCHIEHLEN, W., Arcisstraße 21, 8000 München 2	Germany

SHARMA, S.D.,
Lämmersieth 90, 2000 Hamburg 60 Germany

VAN SWIETEN, A.,
P.O.Box 800, Groningen Netherlands

TESCHNER, W.,
An der Steinkaute, 6109 Mühltal 4 Germany

VINCENT, Th.L.,
Univ.of Western Austr., Nedlands.W.A.6009 Australia

VRIEZE, K.,
2e Boerhaavestraat 49, Amsterdam Netherlands

VAN DER WAL, J.,
P.O.Box 513, Eindhoven Netherlands

WANROOY, G.,
2e Boerhaavestraat 49, Amsterdam Netherlands

WILLEMS, J.C.,
P.O.Box 800, Groningen Netherlands

ZINTERHOF, P.,
Petersbrunnstraße 19, A-5020 Salzburg Austria

LIST OF SHORT COMMUNICATIONS

BASAR, T.,	- Deterministic nonzero-sum dynamic games and informationally nonunique Nash equilibrium solutions
FLEMING, W.,	- Optimal inclusion probability and differential games
GRONSKI, J.,	- Controllability on compact connected manifolds
LEWIN, J.,	- On estimations of the "Value" of a game
PACHTER, M.,	- Some properties of the value and strategies in infinite-time linear-quadratic differential games
PARTHASARATHY, T.,	- On measurable relations
PIKKEMAAT, G.,	- Income distribution as a dynamic game
REBHUHN, D.,	- Stable singular controls for nonlinear control systems
SACHS, E.,	- A nonlinear parabolic boundary control problem
SHARMA, S.,	- Maritime collision avoidance as a differential game
VAN SWIETEN, A.,	- Linear-quadratic zero sum differential games
WILLEMS, J.,	- Linear-quadratic pursuit-evasion games

Introduction

Historically, early works in differential games and optimal control theory appeared almost simultaneously, independently of each other. The theory of differential games, however, got a big boost immediately after the establishment of optimal control theory in the late fifties. This succession, at first glance, seems to be natural, since a differential game may be viewed upon as a control process where the controls are divided among various players who are willing to use them for objectives which possibly conflict with each other. A closer look however will reveal that the development of the two fields followed different lines. Both have the dynamic aspect in common, but differential games have in addition a game-theoretic aspect. The latter overshadows so much the dynamics that the techniques developed in optimal control theory cannot simply be reused in differential games. This basic difference is the reason why the picture of the present situation in differential games appears to the casual observer rather unfavorable compared with control theory, and he feels disturbed by the fact that in differential games mathematical rigor and geometric intuition have still not blended into a satisfactory theory. This impression has two negative effects. First it discourages workers in applied sciences, who could use differential games, to make an effort in learning the background. Secondly the subject is not adequately represented in the curriculum of many universities and Institutes of Technology, hence young engineers and mathematicians rarely have the opportunity to acquire even a modest basic knowledge of the subject.

It was the concern about this situation and the desire to seek ways of improving it which gave rise to the organization of the workshop. The basic idea of the workshop was that it should be accessible to all people with interest (but not specific background) in differential games, and that it should provide the audience with a more accurate picture of the present state of the art and in particular of the scope of the methods. The success of the workshop therefore depended very much upon the cooperation of listeners and speakers; in particular, upon the willingness of the speakers to make the objectives of the advanced parts of their lecture understandable to a broader audience. It should be gratefully acknowledged that most of the speakers did their best to meet this request. In fact some spent considerable time outside the regular lecture hours to explain with patience the basic concepts of differential games.

A list of participants has been included in this volume and it shows that twelve countries from all over the world were represented. Participants have been encouraged to present informal communications along with the invited lectures and twelve participants availed themselves of this opportunity.

This volume contains all invited lectures presented at the workshop. The material canbe divided into four groups.

1.) The largest group (Bernhard, Blaquiere, Breakwell, Leitmann) represents what may be viewed as the continuation of Isaac's work. It provides an informative cross-section about the tools which are presently available in order to attack concrete problems in pursuit and evasions games, zero-sum games, cooperative and other types of dynamical games.

2.) The second group (Elliott, Hajek) is devoted to basic questions. It illustrates some of the present efforts to deal with existence problems and to clarify the concepts of solutions and strategies.

3.) The third group (Olsder, Roxin) concerns problems not described by ordinary differential equations and indicates present activities in neighbouring fields.

4.) The fourth group (Case, Vincent) contains a selection of unconventional problems which arise in applications. These papers illustrate how attractive dynamic games can be for those who like the

straightforward and intuitively motivated approach.
We hope that the publication of the abstracts will contribute toward the general aims of the workshop: to stimulate interest in the subject and to provide information to all who feel that differential games could be useful for their own research or should have a place in the teaching program of their own university. The support of the Stiftung Volkswagenwerk, which provided the financial basis for the workshop, and the hospitality of the University of Twente is gratefully acknowledged.

Darmstadt, May 1977 Hagedorn, Knobloch, Olsder

Singular Surfaces in Differential Games

An introduction

Pierre BERNHARD
Université de Paris IX

and

Centre d'Automatique et Informatique
de l'Ecole Nationale Supérieure
des Mines de Paris

Abstract.

We give a general set up and a version of Isaacs' Verification Theorem that allows us to deal with the various singularities we want to investigate. In particular, we are obliged to allow upper or lower strategies, leading to upper or lower saddle points, that may exists even if the Hamiltonian does not have a saddle point. It is shown that this is needed even for separated games. Then we give a general study of junctions of optimal fields with singular surfaces, which requires a special investigation of the situation where this junction is tangantial extending Carathéodory's General Envelope Theorem. We then proceed to study special singular surfaces, and we end up with an example which shows how a state constraint may appear in the interior of the game space of a separated problem posed with no such constraint to start with.

Introduction

It can well be said that Isaacs founding work on two person, zero sum, Differential Games, [1] is mainly a study of singular surfaces (together with the fundamentals of Hamilton Jacobi theory). While this topic was investigated further by J.V. Breakwell and his students in particular among other works, see [2], most of the following work has been in the area of existence theory, by refining the concept of strategy (See, e.g. [3] to [11].) While this later work is highly significant, and of relevance to the present one, we wish here to turn back to the topic of better understanding singular surfaces. For the sake of brevity, we shall omit topics related to existence theory, in particular the justification of our choice of upper and lower strategies, and the question of their approximation by simple strategies.

The main emphasis of this short course will be on giving a more rigorous treatment of long used practices, particularily in the case where optimal trajectories have an envelope, a situation that has been recognized early by J.V Breakwell in the study of particular games [12], [13]. In the process, we hope to contribute to a unification, thus a simplification, of the whole topic.

In part I, we shall give the general set up we use, and the relevant Hamilton Jacobi Isaacs equation for this set up. In part 2 we give the fundamental lemmas that allow us to deal with the envelope situation, and general results on junctions of an optimal field with a singular surface. In part 3 we investigate various kinds of singularities Part 4 gives a simple and interesting example : the One Dimensional Second Order Servomechanism Problem.

1. General set up

1.1. The Game

We shall consider a two player dynamical system governed by the differential equation

(1) $\dot{x} = f(x, u, v)$

where $\dot{x}$ means dx/dt, $t \in R^+$ is the time,

$x(t)$ is the state, $x(t) \in X \subset R^n$
$u(t)$ is the first player's control, $u(t) \in \mathcal{U} \subset R^l$, $\mathcal{U}$ closed,
$v(t)$ is the second player's control, $v(t) \in \mathcal{V} \subset R^m$, $\mathcal{V}$ closed.

Admissible control functions are

$u(.) \in \Omega_u$ = piecewise continous functions fromm R^+ into $\mathcal{U}$
$v(.) \in \Omega_v$ = piecewise continous functions from R^+ into $\mathcal{V}$

$f(., ., .)$ is a C^2 function fromm $R^n \times R^1 \times R^m$ into R^n. We shall assume that f is such that for every pair of admissible controls $u(.)$, $v(.)$, (1) has a bounded solution on every finite interval of the positive real line from any intial state x_o in X. Some further assumptions on f, U and V will be made in a moment.

Notice that we have taken an autonomous system. It is a well known fact that it is always possible to do so, if necessary by having the last component of x have a constant unit time derivative, and thus be equal to time. This also allows us to take, thereafter, an autonomous definition of payoff and strategies, and to assume that the game always begin at time zero.

We could have generalized slightly by allowing measurable control functions, or, more significantly, by allowing $\mathcal{U}$ and $\mathcal{V}$ to vary with time alone, or with all of x in an upper semi-continuous fashion. We avoid it here for the sake of simplicity. However, since x may contain time as one component, taking X fixed as we shall do does not imply that the actual capture set (or capture zone boundaries such as barriers) is (are) fixed.

The <u>playing space</u>. X will be assumed to be a closed subset of R^n, with non empty interior, locally on one side of its C^2 boundary ∂X.

<u>Final time</u> t_1 is the last instant of time before $x(t)$ leaves X:

$$t_1 = \sup \{ t \in R \mid \forall \tau \in [0,t], x(\tau) \in X\}.$$

t_1 is a function of initial state and of the chosen control functions $u(.)$ and $v(.)$. It may be infinite.

A <u>pay off</u> is associated to each initial state and pair of control functions :

$$J(x_o ; u(.), v(.)) = K(x(t_1)) + \int_0^{t_1} L(x, u, v)\, dt \quad \text{if } t_1 < \infty$$

$$J(x_o ; u(.), v(.)) = +\infty \quad \text{if } t_1 = \infty$$

(Therefore we arbitrarily decided that the minimizing player wants the game to terminate).

We shall assume that, for all x in X, and v in $\mathcal{V}$, the set $(L(x, \mathcal{u}, v), f(x, \mathcal{u}, v))$ is convex and bounded in R^{n+1}, and similarily in $\mathcal{V}$. In fact, the only property we need is that for relevant vectors λ or R^n, $L + \lambda' f$ have a unique minimum in u or maximum in v.

$K(x)$ is a function from R^n into $R \cup \{-\infty, +\infty\}$. That is, a part ∂X_u of ∂X may exist such that

$$J = K(x(t_1)) = +\infty \qquad \text{if} \quad x'(t_1) \in \partial X_u$$

and similary a part ∂X_v where J is equal to $-\infty$ if the game terminates there. We say that ∂X_u defines a state constraint under the first players responsability and similarily for ∂X_v. $K(.)$ is assumed to be of class C^2 in the interior of the region where it is finite.

The first player, or Pursuer P wants to minimize J while the second player, or Evader E wants to maximize it. However this statement must now be made more precise by specifying the information available to the players in making their choice (strategy concept) and the solution sought.

1.2. Strategies and saddle points.

Using an idea of Varaiya and Roxin [5], [6], [7], and a special form of Isaacs tenet of transition (which he stated as early as 1952 in Rand seminars), one can justify the following definitions, that we shall take here as part of the statement of the game.

A u - discriminating strategy or u - D - strategy, (the need for this type of strategy was probably first seen by J.V. Breakwell [14]) is an application $\varphi \in \Phi^-$ from $X \times \mathcal{V}$ into $\mathcal{u}$:

(3a) $$u = \varphi(x, v)$$

such that for every admissible control function $v(.) \in \Omega_v$, the differential equation

(4a) $$\dot{x} = f(x, \varphi(x, v), v) \qquad x(o) = x_o ,$$

has a unique solution for every initial state x_o in X, in the precise following meaning : there exists an absolutely continuous function $x(t)$ satisfying (4a) for each t for which $x(.)$

is differentiable (i.e. allmost all t), that together with $v(.)$ generates via (3a) an admissible control function $u(.) \in \Omega_u$. This defines the set Φ^- of admissible u-D-strategies.

A <u>v-ordinary strategy</u> (or v-strategy) is an application $\psi \in \Psi$ from X into $\mathcal{U}$:

(3b) $$v = \psi(x)$$

such that for every admissible control function $u(.) \in \Omega_u$, the differential equation

$$\dot{x} = f(x, u, \psi(x)) , \qquad x(0) = x_0 ,$$

has a unique solution for every initial state x_0 in X, with the same definition as above, that, together with $u(.)$, generates via (3b) an admissible control function $v(.) \in \Omega_v$. This defines the set Ψ of admissible v-strategies.

A <u>lower saddle point</u> is a pair (φ^-, ψ^-) of admissible u-D-strategy and v-strategy such that

i) the differential equation

(4c) $$\dot{x} = f(x, \varphi^-(x, \psi^-(x)), \psi^-(x)), \qquad x(0) = x_0,$$

has a solution generating admissible control functions $u(.)$ and $v(.)$. (It suffices to assume that one is admissible, because then the other is such).

ii) for every initial state $x_0 \in X$, there exists a number $V^-(x_0)$ such that, for every admissible control functions $u(.) \in \Omega_u$ and $v(.) \in \Omega_v$,

(5a) $$J(x_0 ; \varphi^-, v(.)) \leq V^-(x_0) \leq J (x_0 ; u(.), \psi^-)$$

The notations $J(x_0 ; \varphi^-, v(.))$ and $J(x_0 ; u(.), \psi^-)$ having an obvious non ambiguous meaning.

Let $u^-(.)$ and $v^-(.)$ be the control functions generated by a solution of (4c) then necessarily

(5b) $$J(x_0 ; u^-(.), v^-(.)) = V^-(x_0)$$

because

$$J(x_o ; u^-(.), v^-(.)) = J(x_o ; \varphi^-, v^-) = J(x_o ; u^-, \psi^-),$$

which together with (5a) implies (5b).

V^- is called the lower value, or lower value function, of the game. Our definition (5a) of a lower saddle point seems restrictive in that it requires comparison controls to be open loop. However, as was pointed out by Berkovitz [15], this is not so since for any closed loop v-strategy ψ, if the pair (φ^-, ψ) generates a solution $x(t)$ admissible in the sense that $\psi(x(.))$ is admissible, then letting $v(.) = \psi(x(.))$ gives the same payoff and allows us to use (5a) to evaluate this payoff. On the other hand, our definition avoids some difficult problems of play ability which may end up in the fact that different saddle points exist for the same game, with different values. [16].
Notice also that since $x(t)$ may have the time as one component our strategies include open loop controls.
We similary define a u-ordinary strategy as an application $\varphi \in \Phi$ of X into $\mathcal{U}$ and a $\mathcal{V}$- D- strategy as an application $\psi \in \Psi^+$ from $X \times \mathcal{U}$ into $\mathcal{V}$. An upper saddle point is a pair $(\varphi^+, \psi^+) \in \Phi \times \Psi^+$, again such that the corresponding differential equation has a solution generating admissible control functions, and such that for any admissible control functions $u(.)$ and $v(.)$, the inequalities (5a) are satisfied with φ^-, ψ^- and V^- replaced by φ^+, ψ^+ and V^+. The upper value V^+ satisfies the equivalent of equality (5b).

We define an ordinary saddle point, or daddle point, as a pair $(\varphi^o, \psi^o) \in \Phi \times \Psi$ of admissible ordinary strategies, with the same properties as before, and such that (5a) hold with φ^o, ψ^o, and V in place of φ^-, ψ^- and V^-.
In most examples, the game shall be "separated", i.e., we shall have

$$f(x, u, v) = g(x, u) + h(x, v)$$

$$L(x, u, v) = M(x, u) + N(x, v)$$

In these cases, the hamiltonian (that we introduce below) has a saddle point in (u, v), and it turns out that, except on the singular surfacesWe want to investigate, the optimal D-strategies wont use the extra information allowed on the opponents current control, but will only use the current state. Moreover, on the singular surfaces, only one of the players will need this extra information in order for us to be able to exhibit a solution of the game.

For this reason, it shall then make sense to introduce, without further precautions pertaining to existence, the concept of a D-saddle point where both players are allowed to use D-strategies. A D-saddle point shall be made of the concatenation (in time) of upper and lower strategy pairs, admissible in the same sense as previously, leading to inequalities of the type (5a).

Of course, we do not imply that all these saddle points, or any of them, exist. However, we shall investigate the case where one exists, since we use a theory of sufficiency conditions, the theory of necessary conditions being extremely involved and closely linked to existence theory.

1.3 Isaacs Main equation.

We shall now adapt Isaacs Verification Theorem [1] to the case of a lower saddle point with discontinuities of the Value function.

We introduce a function $V^-(x)$ which is allowed to have discontinuities of a simple kind. (We shall relax this later on). We assume that there exists a partition of X by C^2 n-1-dimensional manifolds, such that

i) $V^-(x)$ is of class C^2 in the interior of each region

ii) its restrictions to these manifolds is C^2

iii) $V(x)$ is continuous and continuously differentiable on at least one side of each manifold, which means that in at least one of the regions, V^- coincide with a continuously differentiable function defined on an open set containing the manifold.

iv) Upon leaving one of these manifolds on a discontinuous side, V^- has a simple jump. We will refer to discontinuity manifolds as being of positive or negative jump according to the sign of this jump (Upon reaching such a manifold on the discontinuous side, the jump is of opposite sign).

Remark that for the following theorem, it suffices to assume for $V^-(x)$ the regularity C^1 where we have assumed C^2. However, since we shall construct solutions making use of the characteristics of Isaacs equation (his retrograd path equations or Euler

Lagrange equations) we shall not make use of any more generality. This is not so for the point iii) which is precisely the hypothesis we even want to relax further later on.

Notice also that at the intersection of discontinuity manifolds there exists manifolds of lower dimension where both positive and negative jumps may occur. In our local investigation of discontinuity manifolds, we shall not consider these higher order singularities. For the global theorem we state in this paragraph, this only translates into intersections of the set we now proceed to define, and causes no special problem.

For each $x \in X$ and $v \in \mathcal{V}$, let $\tilde{\mathcal{U}}_v(x)$ be the set of controls u that prevent a positve jump, that is

$\tilde{\mathcal{U}}_v(x) = \{u \in \mathcal{U} | \nu'(x) f(x, u, v) \leq 0\}$ if $x \in$ positive jump manifold, with normal $\nu(x)$ pointing toward the discontinuity.

$\tilde{\mathcal{U}}_v(x) = \mathcal{U}$ otherwise.

Similarily, let $\tilde{\mathcal{V}}(x)$ be the set of controls $v \in \mathcal{V}$ that prevent a negativd jump <u>for all</u> $u \in \mathcal{U}$.

We shall hereafter assume that the property $u(t) \in \tilde{\mathcal{U}}_{v(t)}(x(t))$ for all t implies that the trajectory does not leave a positive jump discontinuity manifold on the discontinuous side. This is not rigorously true, some more care is required, but we shall not go into this question in any more detail. Similarily for $v \in \tilde{\mathcal{V}}(x)$ and negative jump manifolds.

In order to state our theorem, we introduce the hamiltonian of the game

$$H(x, \lambda, u, v) = L(x, u, v) + \lambda' f(x, u, v)$$

which is a function from $X \times R^n \times \mathcal{U} \times \mathcal{V}$ into R.

THEOREM 1. Assume there exists a function $V^-(x)$ defined over X, with the regularity described above, and a lower strategy pair $(\varphi^-, \psi^-) \in \Phi^- \times \Psi$ such that

i) for every admissible control function $v(.) \in \Omega_v$, trajectories generated by $(\varphi^-, v(.))$ never reach a negative jump manifold from the discontinuous side, and

(6a) $$\forall\, x \in X, \quad \forall\, v \in \mathcal{V}, \qquad \varphi^-(x, v) \in \tilde{\mathcal{U}}_v(x)$$

ii) for every initial state x_0 in X and every admissible control function $v(.) \in \Omega_v$, the game terminates at a finite time t_1 (1).

iii) for every admissible control function $u(.) \in \Omega_u$, trajectories generated by $(u(.), \psi^-)$ never reach a positive jump manifold from the discontinuous side, and

(6b) $$\forall\, x \in X, \quad \psi^-(x) \in \tilde{\mathcal{V}}(x)$$

iv) ∂X is treated as a (possible) discontinuity manifold, with $V^-(x) = K(x)$ in the exterior of X (Thus ∂X is a positive jump manifold where $V^-(x) < K(x)$, and conversely). There necessarily exists a region of ∂X, called the usable part, where $V^-(x) = K(x)$, otherwise ii) could not hold for the trajectory φ^-, ψ^-.

v) The following relations hold everywhere in X :

(7a) $$H(x, \frac{\partial V^-}{\partial x}(x), \varphi^-(x, \psi^-(x)), \psi^-(x)) = 0,$$

(7b) $$\forall\, v \in \tilde{V}(x), \qquad H(x, \frac{\partial V^-}{\partial x}, \varphi^-(x, v), v) \leq 0,$$

(7c) $$\forall\, u \in \tilde{\mathcal{U}}_{\psi^-(x)}, \quad H(x, \frac{\partial V^-}{\partial x}, u, \psi^-(x)) \geq 0.$$

Then, (φ^-, ψ^-) is a lower saddle point and $V^-(x)$ the associated lower value.

PROOF

Assumption i) and iii) imply that a trajectory generated by (φ^-, ψ^-) has no jump in $V(x(t))$. Therefore, noticing that

$$H(x, \frac{\partial V^-}{\partial x}, u, v) = \frac{dV^-}{dt}(x, u, v) + L(x, u, v)$$

where the total time derivative is taken along the trajectory, (7a)

---

(1) We assume that the game of kind has been dealt with before and that X is the capture region from where P is able to force the game to terminate. ∂X may contain barriers, that form part of ∂X_u.

yields for a trajectory generated by (φ^-, ψ^-)

$$V^-(x(t_1)) + \int_0^{t_1} L(x, u, v)\, dt = V^-(x_0)$$

(Because of ii), t_1 exists). Because of iv) and relations (6);

$$V^-(x(t_1)) = K(x(t_1))$$

so that the previous relation yields

$$J(x_0 \,;\, \varphi^-, \psi^-) = V^-(x_0)$$

which is relation (5b).
Now, consider an arbitrary admissible control function $v(.)$, and the trajectory generated, from a prescribed initial phase x_0, by the pair $(\varphi^-, v(.))$. By assumption it terminates at a finite time that we again note t_1. Because, of i), any jump in $V^-(x(t))$ will be negative. However, as V^- remains finite, there shall be at most countably many such jumps. At these instants, $v \notin \tilde{V}(x)$. However, (7b) will still hold for almost all t, or :

$$\frac{dV}{dt}(x, \varphi^-, v, t) + L(x, \varphi^-, v, t) \leq 0 \quad \text{almost all } t.$$

We can therefore integrate, and we get (from the sign of the jumps) :

$$V^-(x(t_1)) - V^-(x_0) \leq \int_0^{t_1} \frac{dV^-}{dt} \leq -\int_0^{t_1} L(x, u, v)\; dt$$

Now, from iv) and (6 a) results that

$$V^-(x(t_1)) \leq K\,(x(t_1))$$

so that we get

$$J(x_0 \,;\, \varphi^-, v(.)) \leq V^-(x_0)$$

Finally, consider an arbitrary admissible control function $u(.)$, and the trajectory generated by $(u(.), \psi^-)$. Either it does not terminate, then $J = +\infty$, either it does terminate and we have a similar argument using (7c). In both cases we conclude that

$$J(x_0 \,;\, u(.), \psi^-) \geq V^-(x_0)$$

and the proof is complete, since the last two in equalities are identical to (5a).

REMARK. While (7a) and (7b) imply that $\psi^-(x)$ yields the maximum over $\tilde{\mathcal{V}}(x)$ of $H(x, V_x^-, \varphi^-(x, v), v)$, (7c) does not imply that $\varphi^-(x, v)$ yields the minimum over $\tilde{u}_v(x)$ of $H(x, V_x^-, u, v)$ for all v, but only for $v = \psi^-(x)$. However, since

$$\min_{u\in\tilde{u}_v} H(x, \frac{\partial V^-}{\partial x}, u, v) \leq H(x, \frac{\partial V^-}{\partial x}, \varphi^-(x, v), v) \leq 0$$

The argument of the minimum then satisfies all the hypothesis in the theorem, except perhaps ii). In any event, condition v) implies

$$\max_{v\in\tilde{\mathcal{V}}(x)} \min_{u\in\tilde{u}_v(x)} H(x, \frac{\partial V^-}{\partial X}, u, v) = H(x, \frac{\partial V^-}{\partial x}, \varphi^-, \psi^-) = 0$$

The rather complicated set up used here is devised to allow an optimal trajectory to reach the boundary of the capture set, for instance, or another particular manifold such as a barrier, and possibly stay on it for a while. It also takes care, through the requirement (6a) of the case where the game starts from a barrier limiting the capture zone and where by playing the barrier strategy, E may prevent P from using his ordinary saddle point strategy (Breakwell's lunge maneuver. See [17],)

We just noticed that we dot need that H have a saddle point. Our examples shall be <u>separated games</u>, i.e. games where f and L are such that we have

$$H(x, \lambda, u, v) = H_P(x, \lambda, u) + H_E(x, \lambda, v)$$

For such games as we said , H as a saddle point and optimal D-strategies turn out to be ordinary strategies, except, and this is very important, on discontinuity manifolds because the requirement (6a) couples u and v. It is why separated games may not have ordinary saddle points. The example of part 5 is an instance of this fact. (See [18], [19] for special uses of upper saddle point).

2. Junction with a singular surface.

We shall show later on that in general, when an optimal trajectory reaches a singular manifold, the situation we wish to allow, it does so tangentially. For reasons we shall make clear in a moment most of the classical litterature rules out this situation. The aim of this part is to show that we can deal with this case as with the non tangential case.

2.1. Non differentiability of a continuous V

The following developpement holds for any kind of saddle point. We note (φ^*, ψ^*) the optimal strategies, and $V(x)$ the correspon-ding value, and to investigate these junctions."

$$f^*(x) = f(x, \varphi^*, \psi^*), \qquad L^*(x) = L(x, \varphi^*, \psi^*)$$

Let S be a $n-1$ dimensional manifold locally parametised by the C^2 map

$$x = \xi(s) \ , \qquad s \in \theta \subset R^{n-1}$$

By assumption, the restriction of $V(x)$ to S is a C^2 function :

$$V(\xi(s)) = U(s)$$

A field of optimal trajectories reaches S. Let $t = \tau(s)$ be the time at which the trajectory through $\xi(s)$ reaches S.
We have, along that trajectory, and for $t < \tau(s)$.

$$(8) \qquad x(t) = \xi(s) + \int_{\tau(s)}^{t} f^*(x(\alpha))d\alpha \triangleq y(s,t)$$

By assumption, this field is regular, in the precise meaning that $(\partial y/\partial s, \partial y/\partial t)$ exists is bounded in the closed half space considered, and is invertible in the open half space (to allow a tangent contact). Notice that :

$$\frac{\partial y}{\partial t} = f^*(y)$$

Finally, the optimal trajectories defines a value function by

$$(9) \qquad V(y(s,t)) = U(s) + \int_{t}^{\tau(s)} L^*(y(s,t))\, dt \triangleq W(s,t)$$

This last relation gives, in matrix notations

$$(10) \quad \frac{\partial V}{\partial x}\left(\frac{\partial y}{\partial s}\ \frac{\partial y}{\partial t}\right) = \left(\frac{\partial W}{\partial s}\ \frac{\partial W}{\partial t}\right).$$

Therefore, at every point where the inverse exists

$$\frac{\partial V}{\partial x} = \left(\frac{\partial W}{\partial s}\ \frac{\partial W}{\partial t}\right)\left(\frac{\partial y}{\partial s}\ \frac{\partial y}{\partial t}\right)^{-1}$$

However, if the optimal trajectories reach S tangentially, this implies that $\partial y/\partial t$, which is equal to f^*, is linearly dependent with $\partial\xi/\partial s$ which is, by definition, the set of (column) vectors generating the tangent plane to S. In this case, $\partial V/\partial x$ need not exist, and usually does not.

It is interesting to see what (10) gives at $t = \tau(s)$ when $\partial V/\partial x$ exists there. Assuming f^* defines a regular field on S (i.e., $\partial f^*/\partial s$ exists), we differentiate (8) and (9) partially at $t = \tau(s)$ it comes, usign $y(s, \tau(s)) = \xi(s)$:

$$\frac{\partial y}{\partial s}(s, \tau(s)) = \frac{\partial \xi}{\partial s} - f^*(\xi(s))$$

$$\frac{\partial W}{\partial s}(s, \tau(s)) = \frac{\partial U}{\partial s} + L^*(\xi(s))$$

Now, the last column of (10) gives (for all t)

$$\frac{\partial V}{\partial x} f^*(x) + L^*(x) = 0$$

(which is (7a)), and this together with the first block column of (10) gives

$$\text{(11a)} \qquad \frac{\partial V}{\partial x}(\xi(s))\,\frac{\partial \xi}{\partial s} = \frac{\partial U}{\partial s}$$

which is the classical fact that the gradient of V has its projection on the tangent plane to S equal to the gradient of the restriction U of V to S.

2.2. The envelope lemma.

We now consider the case of the tangent field. S is then the enveloppe of this field. This means that along a trajectory, φ^* and ψ^* are continuous functions of time with, $f^*(y(s, \tau(s))) \in$ tangent plane to S. We make an assumption of regularity on the field near the contact :

ASSUMPTION. The direction f^* of the optimal trajectories has, as a function of s and t, a continuous partial derivative $\partial f^*/\partial s$ in the closed half space.

REMARK We specifically avoid to assume that $\partial f^*/\partial x$ exists and is continuous, since φ^* and ψ^* usually depend on the gradient V_x. Our assumption is that the field of optimal directions in S is regular, and varies smoothly in the neighborhood of S.

LEMMA. Under the above conditions, the gradient of V has a limit λ_o ax $x \to S$ on an optimal trajectory, and this limit satisfies the relation

$$\lambda_o' \frac{\partial \xi}{\partial s} = \frac{\partial U}{\partial s} \tag{11b}$$

PROOF. In the open half space, $\tilde{u}_\psi = u$ and $\tilde{\nu} = \nu$. Thus, using standard techniques of control theory, we have

$$\frac{\partial \overline{H}}{\partial x} = \frac{\partial H}{\partial x}\left(x, \frac{\partial v}{\partial x}, \varphi^*(x), \psi^*(x)\right),$$

where $\overline{H}(x, \lambda) = H(x, \lambda, \varphi^*(x), \psi^*(x))$.

Therefore, if λ has a limit as $x \to S$, so does $\partial\overline{H}/\partial x$.

In the open half space, let $\lambda = (\partial V/\partial x)'$ be the gradient of V. V being assumed to be of class C^2, λ satisfies the Euler Lagrange equation :

$$\dot{\lambda}' = -\frac{\partial \overline{H}}{\partial x} = -\frac{\partial H}{\partial x} = -\lambda' \frac{\partial f}{\partial x}(x, \varphi^*, \psi^*) - \frac{\partial L}{\partial x}(x, \varphi^*, \psi^*).$$

This is a linear differential equation in λ with bounded coefficients. Therefore λ remains finite as $t \to \tau(s)$, and has a limit λ'_o given as a function of λ at a previous time t : by

$$\lambda'_o(s) = \lambda'(t) - \int_t^{\tau(s)} \frac{\partial H}{\partial x}(y(s, \alpha))\, d\alpha.$$

(with a transparent abuse of notation).

Consider a particular s and $t < \tau(s)$. Then by assumption $\partial f^*/\partial s$

exists from t to τ, and thus also $\partial L^*/\partial s$. Differentiate partially (8) and (9) and place in (10). The first block column yields :

$$\frac{\partial V}{\partial x}\left(\frac{\partial \xi}{\partial s} + \int_{\tau(s)}^{t} \frac{\partial f^*}{\partial s}(y(s,\ \alpha))d\alpha - f^*(\xi(s))\frac{d\tau}{ds}\right) - \frac{\partial U}{\partial s}$$
$$+ \int_{\tau(s)}^{t} \frac{\partial L^*}{\partial s}(y(s,\alpha))d\alpha - L^*(\xi(s))\ \frac{d\tau}{ds} = 0.$$

As previously, we use tha last column of (10) to cancel the terms multiplying $d\tau/ds$, and using (12) again, we get

$$\frac{\partial V}{\partial x}\ \frac{\partial \xi}{\partial s} - \frac{\partial U}{\partial s} + \int_{\tau(s)}^{t} \frac{\partial H}{\partial x}\ \frac{\partial y}{\partial s}\ d\alpha = 0.$$

We take the limit as $t \uparrow \tau(s)$. By assumption the integrand remains bounded since

$$\frac{\partial y}{\partial s}\ (\alpha) = \frac{\partial y}{\partial s}\ (t) + \int_{t}^{\alpha} \frac{\partial f^*}{\partial s}\ (\beta)\ d\beta,$$

and we obtain the result sought.

COROLLARY. The theorem 1 still holds if V has the type of singularity we have described above.

PROOF. We just have to check that for all trajectories avoiding a jump,

$$V(x(t_1)) - V(x(t_o)) = \int_{t_o}^{t_1} \lambda'(t)\ \ dt,$$

where $\lambda'(t)$ is either $\partial V/\partial x$ where it exists, or λ'_o on an envelope. If an arc of trajectory lies on S, then this is a consequence of 11b). If an arc has a point $x(t_2)$ not belonging to S, then, by continuity, this is true in an open interval of time. If at an end point t_3 of this interval, $x(t_3)$ belongs to S, then

$$\forall\ t \in (t_2,\ t_3)\ ,\qquad V(x(t)) - V(x(t_2)) = \int_{t_2}^{t} \lambda'(t)\ \ dt$$

and as V is continuous at S, we take the limit of both sides and have the result.

Thus we have extended theorem 1 to a very common type of discontinuity, broadening the applicability of Hamilton Jacobi Isaacs theory, as compared to previous papers attempting to apply it to state constraints for instance.

2.3. Generalized Euler Lagrange equations.

We now need a differential equation for λ_o along an otpimal trajectory lying on S. We have assumed that U(s) is of class C^2. We shall further assume that λ_o is of class C^1 in s.

Let H_o (resp $\overline{H}_o$) be H (resp $\overline{H}$) with λ replaced by λ_o. Since $\overline{H}(x) = 0$ for $t < \tau$, we have in the limit $\overline{H}_o = 0$. Also, (12) gives, in the limit, differentiating $\overline{H}_o$ partially with respect to s_i :

$$\frac{\partial H_o}{\partial x}(x, \lambda_o, \varphi^*, \psi^*) \frac{\partial \xi}{\partial s_i} + \frac{\partial \lambda_o'}{\partial s_i} f^* = 0.$$

Now, along such a trajectory, we have

$$f^* = \frac{\partial \xi}{\partial s}\frac{ds}{dt} = \sum_{j=1}^{n-1} \frac{\partial \xi}{\partial s_j}\frac{ds_j}{dt}.$$

Further, from (11,b) we derive

$$\frac{\partial \lambda_o'}{\partial s_i}\frac{\partial \xi}{\partial s_j} = \frac{\partial^2 U}{\partial s_i \partial s_j} = \frac{\partial \lambda_o'}{\partial s_j}\frac{\partial \xi}{\partial s_i}.$$

Placing the last two equations in the previous one, and using the fact that the scalar ds_j/dt commutes with vectors, we obtain

$$(12a) \qquad \left(\frac{d\lambda_o'}{dt} + \frac{\partial H_o}{\partial x}\right)\frac{\partial \xi}{\partial s_i} = 0, \qquad \forall i.$$

This is a first form of the relation sought. Noticing that $\partial\xi/\partial s_i$ generate the tangent plane to S, we can rewrite it in the following form, where $\nu(\xi)$ is a normal to S and $\alpha(t)$ an unknown scalar function :

$$(12b) \qquad \dot{\lambda}_o' = -\frac{\partial H_o}{\partial x} + \alpha\nu$$

If S is given by an equation $S(x) = 0$, then one possible choice of $\nu(x)$ is $\partial S/\partial x$, and (12b) has a familiar form.

Remark that equations (12) have been established only as a consequence of (7a) and (11b). If a field of optimal trajectories reaches S transversally and then follows S, (with a discontinuity of f^* upon reaching S), then $\partial V/\partial x$ is defined in the closed half space, (11a) replaces (11b), and our generalized Euler Lagrange equations hold.

2.4. <u>Conditions at the junction.</u>

Still in the context of a field of optimal trajectories reaching and then traversing a surface S, and generating a function V satisfying the enlarged theorem 1, we investigate the behaviour of the optimal strategies at the junction with S.

Let $\varphi^*(x, v)$ and $\psi^*(x)$ be the limits of the lower saddle point strategies as x reaches S, and $\hat{\varphi}(x, v)$, $\hat{\psi}(x)$ the "traversing" optimal lower strategies on S. Applying (7a) and (7b) with $\hat{\varphi}$, $\hat{\psi}$, and using the fact that $\tilde{\mathcal{V}}(x) = \mathcal{V}$, it comes (we omitt unnecessary arguments on the functions)

$$(13a) \qquad 0 = H(\hat{\varphi}(\hat{\psi}), \hat{\psi}) \geq H(\hat{\varphi}(\psi^*), \psi^*).$$

Now, in the open halfspace, $\tilde{u}_V(x) = \mathcal{U}$, therefore applying (7c) and (7a)

$$(13b) \qquad H(\hat{\varphi}(\psi^*), \psi^*) \geq H(\varphi^*(\psi^*), \psi^*) = 0.$$

Therefore (13a) gives

$$H(\hat{\varphi}(\psi^*), \psi^*) = 0 = \max_{v \in \mathcal{V}} H(\hat{\varphi}(v), v) = H(\hat{\varphi}(\hat{\psi}), \hat{\psi}).$$

We therefore have,

THEOREM 2. If the maximum of $H(\hat{\varphi}(v), v)$ is unique, the optimal strategy ψ is continuous at the junction.

The geometry of the set $(f(\hat{\varphi}(\mathcal{V}), \mathcal{V}), L(\hat{\varphi}(\mathcal{V}), \mathcal{V}))$ is somewhat difficult to investigate. We shall not attempt here to understand better the unicity assumption of Theorem 2.

Relation (13b) gives similarily :

$$H(\hat{\varphi}(\psi^*), \psi^*) = 0 = \min_{u \in \mathcal{U}} H(u, \psi^*) = H(\varphi^*(\psi^*), \psi^*)$$

We therefore have the interesting result :

THEOREM 3. If the minimum of $H(u, \psi^*)$ is unique, then the optimal trajectories reach S tangentially. If furthermore the optimal strategy ψ is continuous (see theorem 2) then so is the optimal D-strategy φ.

PROOF. Under the assumption of theorem 3, we have

$$\hat{\varphi}(\psi^*) = \varphi^*(\psi^*)$$

and thus $\varphi^*(\psi^*) \in \tilde{u}_{\psi^*}$. Therefore the field generated by (φ^*, ψ^*) cannot be transverse to S.

Under our assumption of convexity of $(f(u,v), L(u, v))$, the minimum is guaranteed to be unique if this set is strictly convex. Otherwise, part of its boundary is linear (a hyperplane, or an intersection of), and we have the following situation :

COROLLARY. Under the convexity assumption, if an optimal field of a lower saddle point reaches a singular surface transversally and then follows it, the hamiltonian is linear with respect to at least one component of u, and is singular at the junction.

Theorem 3 explains why we were interested in allowing envelopes in the field of optimal trajectories.

3. Particular singular surfaces.

Here, we shall study in more details the various situations that may arise according to the shape of the field on both sides of the singular surface. We shall not consider surfaces that are left on both sides by the optimal field (dispersal lines), as they pose no particular problem in the present set up. (This may not be so if we do not allow D-strategies). Neither shall we consider surfaces that are the limit of a field, but not reached by its trajectories. Those are barriers and we purposedly avoid them here.

In the three types we consider, there is a regular case and a singular case according to whether conditions of theorem 3 or of the corollary prevail.

In order to give formulas to actually compute singular surfaces of various kinds, we need the following notations :

$\bar{\varphi}(x, \lambda, v)$, $\bar{\psi}(x, \lambda)$: arguments of $\max_{v \in \mathcal{V}} \min_{u \in \mathcal{U}} H(x, \lambda, u, v)$

$\overline{f}(x, \lambda)$, $\overline{H}(x, \lambda)$, $\overline{H}(x, \lambda)$ are obtained by placing $\overline{\varphi}$ for u and $\overline{\psi}$ for v in f and H.

$\tilde{\varphi}(x, \lambda, v)$ argument of the constrained $\min\limits_{u \in \mathcal{U}_v} : H(x, \lambda, u, v)$

$\tilde{f}$ and $\tilde{H}$ are as $\overline{f}$ and $\overline{H}$ but with $\tilde{\varphi}$ in place of $\overline{\varphi}$.

With $\nu(x)$ the normal to the singular surface, we always have

$$\nu'(x)\tilde{f}(x, \lambda) = 0 \tag{14}$$

and we have know that $\tilde{\varphi}(\overline{\psi})$ differs from $\overline{\varphi}(\psi)$ only in the singular case. In that case, also, we have

$$H(x, \lambda, u, v) = a(x, \lambda, u_2, v)\, u_1 + b(x, \lambda, u_2, v)$$

where u_1 is one component of u, and u_2 all the others. The singularity condition reads:

$$\tilde{a}(x, \lambda) = a(x, \lambda, \tilde{\varphi}_2(x, \lambda, \overline{\psi}), \overline{\psi}(x, \lambda)) = 0. \tag{15}$$

3.1 . State constraints.

This is the case where optimal trajectories exist only on one side of S. Actually, a completely similar situation arises if, on the other side, an optimal field leaves S, but that in addition V(x) has a jump on S. Then S has to be a barrier, of course, but it further plays the role of a local state constraint. We shall give below an example of this very interesting phenomenon.

Fig. 1

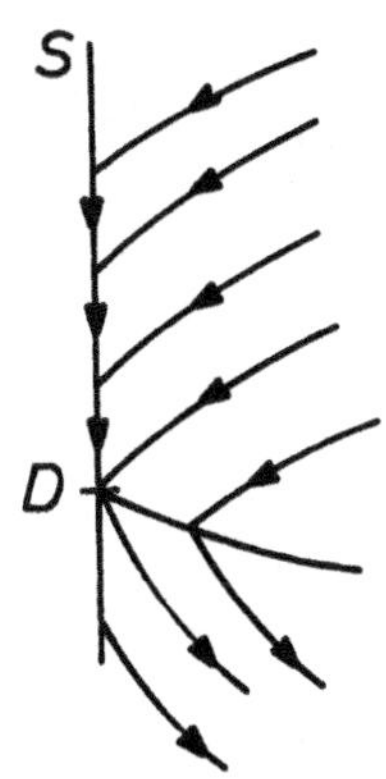

Fig.2

Typically, the field of optimal trajectories constructed backwards hits S along a n-2-dimensional manifold D. Notice that theorem 3 and its corollary apply as well to the point, on D, where an optimal trajectory leaves S. The two typical situations are as depicted in figures (1a) (regular) and (1b) (singular). In the second case, D is the intersection of S with a switching surface for φ^*.

In both cases, the value of λ_0 on the incoming trajectory is assumed known, as this trajectory belongs to the previously computed unconstrained field. Then, the field of optimal trajectories can be computed with the following equations.
Regular case :

$$\dot{x} = \bar{f}(x, \lambda),$$

$$\dot{\lambda} = -\frac{\partial \bar{H}}{\partial x} + \alpha(t) \left(\frac{\partial S}{\partial x}\right)',$$

where α is given by the following equation, obtained by differentiation of

(16) $$\frac{\partial S}{\partial x} \bar{f}(x, \lambda) = 0$$

with respect to time :

$$\alpha \frac{\partial S}{\partial x} \frac{\partial \bar{f}}{\partial \lambda} \left(\frac{\partial S}{\partial x} \right)' + \bar{f}' \frac{\partial S^2}{\partial x^2} \bar{f} + \frac{\partial S}{\partial x} \frac{\partial \bar{f}}{\partial x} \bar{f} - \frac{\partial S}{\partial x} \frac{\partial \bar{f}}{\partial \lambda} \frac{\partial \bar{H}}{\partial x} = 0.$$

Singular case :

$$\dot{x} = \tilde{f}(x, \lambda) ,$$

$$\dot{\lambda} = -\frac{\partial \tilde{H}}{\partial x} + \alpha(t) \left(\frac{\partial S}{\partial x}\right)'$$

Now, (14) is automatically satisfied, but α is chosen in such a way as to insure (15), which, differentiating with respect to time, yields :

$$\alpha \frac{\partial \tilde{a}}{\partial \lambda} \frac{\partial S}{\partial x} - \frac{\partial \tilde{a}}{\partial \lambda} \frac{\partial \tilde{H}}{\partial x} - \frac{\partial \tilde{a}}{\partial x} \tilde{f} = 0.$$

See [20] for a more detailed investigation of state constraints.

3.2. Universal surfaces.

Isaacs called universal surfaces singular surfaces that are reached on both sides by the optimal trajectories.

Let indices 1 and 2 refer to various quantities in the two half spaces, and ν be a normal to S pointing toward region 2. Our lemma shows that :

$$\lambda_1 = \lambda_2 + \alpha\,\nu \tag{17}$$

By assumption, we have, on S

$$\nu' f(\varphi_1^*, \psi_1^*) \geq 0\ , \qquad \nu' f(\varphi_2^*, \psi_2^*) \leq 0.$$

Assume, for instance, that

$$\nu' f(\varphi_1^*, \psi_2^*) \geq 0.$$

By assumption also, both fields satisfy (7), hence,

$$H(\lambda_1, \varphi_1^*, \psi_2^*) = L(\varphi_1^*, \psi_2^*) + (\lambda'_2 + \alpha\nu') f(\varphi_1^*, \psi_2^*) \leq 0$$

$$H(\lambda_2, \varphi_1^*, \psi_2^*) = L(\varphi_1^*, \psi_2^*) + \lambda_2' f(\varphi_1^*, \psi_2^*) \geq 0$$

Hence

$$\alpha\ \nu' f(\varphi_1^*, \psi_2^*) \leq 0, \qquad \alpha \leq 0.$$

Now, looking at $H(\lambda_i, \varphi_2^*, \psi_1^*)$, with $i = 1, 2$, it comes similarily :

$$\alpha\ \nu' f(\varphi_2^*, \psi_1^*) \geq 0$$

and therefore finally

$$\nu' f(\varphi_2^*, \psi_1^*) \leq 0.$$

So, we see that, when the two players chose different strategies, the state always drifts in the half space where E's strategy is optimal. Therefore he can stick to his choice, ψ_1^* in region 1, ψ_2^* in region 2, either of the two on S. P will be unable to keep φ_1^* or φ_2^* since both lead the state in a region where they are not optimal. This situation ends up in a "chatter" for P, or, in our formalism, a strategy φ^* which is not admissible. The only solution for P is to chose, on S, a strategy $\tilde{\varphi}(x, v)$ that will insure that the state remains on S, for all v's. Notice that the situation is different from the state constraint in that, now, P will not let the state drifton either side of S. For that reason, the pairs $(\tilde{\varphi}(x, v), \psi_i^*)$ can be regarded as giving the optimal field of the "reduced game" which is the game where P is obliged to keep the state on x, that is where $u \in \hat{u}_V(x)$ of all controls that satisfy $\nu' f = 0$. Now, except perhaps on exceptional dispersal lines, this game has under our assumptions of convexity, a unique field of extremals. Therefore

$$f(x, \tilde{\varphi}(x, \psi_1^*(x)), \psi_1^*(x)) = f(x, \tilde{\varphi}(x, \psi_2^*(x)), \psi_2^*(x)).$$

(But this does not imply that $\psi_1^* = \psi_2^*$).

Again, two situations arise regular and singular. In the first case, we have A. Merz's focal line [14], [17]. The fields come in tangentially. In the second case there is a corner at the junction. We have the equivalent of singular arcs of optimal control theory. The typical situations are described in figures (2a) and (2b) respectively.

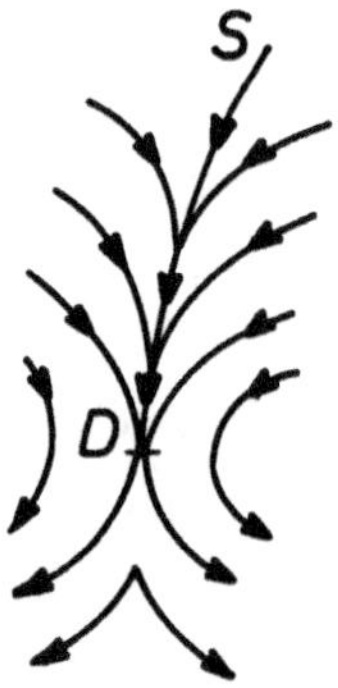

Fig.2a

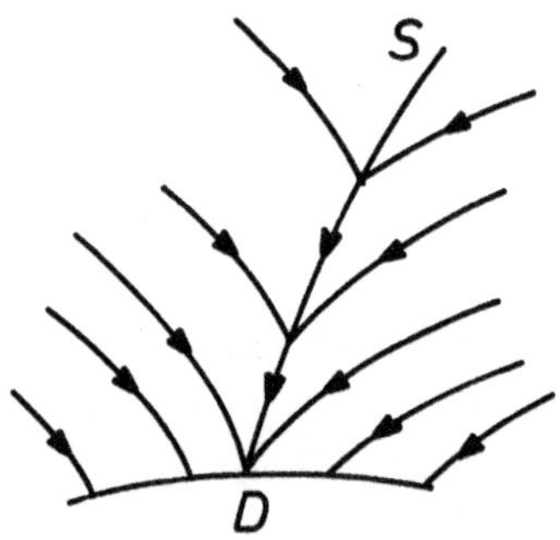

Fig.2b

Here, D has to be found as a locus of points where the trajectories of two fields are tangent, or as a switching point on a surface otherwise joined by the trajectories. Again, λ on the incoming trajectory at D is thus known. (See, however, a more complicated situation in [20]).

In the present case we do not know S a priori, but we know that $\lambda_2 - \lambda_1$ must be normal to it. We can therefore proceed as follows, from D :

Regular case :

$$\dot{x} = \bar{f}(x, \lambda_1) ,$$

$$\dot{\lambda}_1 = - \frac{\partial \bar{H}_1}{\partial x} + \alpha_1(\lambda'_1 - \lambda'_2).$$

$$\dot{\lambda}_2 = - \frac{\partial \bar{H}_2}{\partial x} + \alpha_2(\lambda'_1 - \lambda'_2).$$

$\alpha_1(t)$ and $\alpha_2(t)$ being chosen in such a way that

$$(\lambda'_1 - \lambda'_2)\bar{f}(x, \lambda_1) = 0$$

$$(\lambda'_1 - \lambda'_2)\bar{f}(x, \lambda_2) = 0$$

We leave to the reader to carry out the time differentiations to get a system for α_1 and α_2.

Singular case : same equations as above, except that the bars must be replaced by tildas and $\tilde{u}_v$ is defined using $\lambda_2 - \lambda_1$ for ν. Now, α_1 and α_2 are chosen to insure (15) with both λ_1 and λ_2, and are still explicitely given by differentiation of these two formulas with respect to time.

What remains to be checked at this point is that the above construction actually leads to a hamiltonian that remains null, and, more difficult, to a surface S normal to $(\lambda_1 - \lambda_2)$. To do this, one must parametrize the final conditions on D with a (n-2) dimensional parameter s, and show that the above equations lead to a surface $x(t) = \xi(t,s)$ satisfying :

$$(\lambda'_1 - \lambda'_2) \frac{\partial \xi}{\partial t} = 0, \quad (\lambda'_1 - \lambda'_2) \frac{\partial \xi}{\partial s} = 0$$

The first of these relations will be satisfied by contruction. Checking the second is more difficult. It can be done using the same argument as for the corner surfaces of the next paragraph. The details can be found in [20] and [21].

3.3. Corner surfaces.

Here we are interested in surfaces that are reached by the optimal trajectories on side 1 and left on side 2. Hence, V(x) will again be continuous on both sides, but, as previously will be joined tangentialy in the regular case. ν is again pointing toward region 2.(17) still holds.

By assumption, we have

$$\nu' f(\varphi_1^*, \psi_1^*) \geq 0, \quad \nu' f(\varphi_2^*, \psi_2^*) \geq 0.$$

We distinguish two cases, depending on whether the following <u>permeability condition</u> is met or not :

CONDITION

$$\nu' f(\varphi_1^*, \psi_2^*) > 0, \qquad \nu' f(\varphi_2^*, \psi_1^*) > 0.$$

This condition says that neither of the two players can, by refusing to switch to the strategy 2 upon reaching S, prevent the other from doing so and be right to.

We then have the equivalent of Weierstrass' corner condition :

THEOREM 4. If the permeability condition is met, the gradient must be continuous across S.

PROOF. As previously, we use (7b) and (7c) in both fields using (17) :

$$H(\lambda_1, \varphi_1^*, \psi_2^*) = L(\varphi_1^*, \psi_2^*) + (\lambda_2' + \alpha\nu') f(\varphi_1^*, \psi_2^*) \leq 0$$

$$H(\lambda_2, \varphi_1^*, \psi_2^*) = L(\varphi_1^*, \psi_2^*) + \lambda_2' f(\varphi_1^*, \psi_2^*) \geq 0$$

hence

$$\alpha\nu' f(\varphi_1^*, \psi_2^*) \leq 0$$

Similarily, using (φ_2^*, ψ_1^*), we get

$$\alpha\nu' f(\varphi_2^*, \psi_1^*) \geq 0$$

This together with the permeability condition implies

$$\alpha = 0$$

and the theorem is proved.

Therefore, a corner can occur only if a continuous gradient causes a switch in the optimal controls, which happens with a singular hamiltonian, a classical situation, met in the Dolichobrachistochrone problem for instance [1], and in the example below.

Now, let us assume that

$$\nu' f(\varphi_2^*, \psi_1^*) \leq 0.$$

From the proof of the above theorem, we infer that

$$\alpha \leq 0,$$

$$\nu' f(\varphi_1^*, \psi_2^*) \geq 0.$$

As in the case of the universal surfaces, if E decides to keep his strategy ψ^*, on S, the strategy φ^* is not admissible for P, nor any other that does not keep x on S.
Therefore, against ψ_1^*, P must play a strategy $\tilde{\varphi}$ that makes the state transverse S, and we are in the situation of our general theory of junction of singular surfaces. Notice that here, at any time while traversing S, E may chose to switch to ψ_2^*, and then the state will leave S on side 2. P must then switch to φ_2^*.
The regular case corresponds to J.V. Breakwell's switch envelope [13], [17], and the singular case to R. Isaac's equivocal surface [1]. See [22] or [23] for a more detailed discussion.

Now, as far as constructing S is concerned, the situation is somewhat different. We assume that the field 2 has been constructed previously, and a singularity D is known. It can be a corner of the game of kind, or a more complex situation. (See [20]). The equations used are the same as the first two we wrote above for universal surfaces. A difficulty appears in differentiating the relation

$$(\lambda_1' - \lambda_2')\bar{f}(x, \lambda) = 0,$$

or relation (15) in the singular case, with respect to time. It occurs because $\partial\lambda_2/\partial t$ is not directly known along the trajectories we construct. We must then assume that, in the field 2, we know $\lambda_2 = \partial V/\partial x$ as a function of x, and are able to compute

$$\frac{\partial\lambda_2}{\partial t} = \frac{d\lambda_2}{dx}\,\overline{f}(x,\ \lambda_2)$$

As previously, to have a satisfactory theory, we must still check that S_{so} constructed is actually normal to $(\lambda_1-\lambda_2)$. See [20] for a complete proof.

4. The second order servomechanism problem.

4.1. Statement of the problem

A simple second order plant, with state y, is governed by

$$\ddot{y} = v \qquad v \in \mathcal{V},\ \text{a bounded set.}$$

It is to match a set point z that may drift in an impredictable fashion, but with bounded speed :

$$\dot{z} = u \qquad u \in \mathcal{U},\ \text{a bounded set.}$$

The specification of the servomechanism is its precision :

$$z-y \in \mathcal{X},\quad \text{a bounded set.}$$

We consider the simpler case where y and z are scalars, all data sets are symetrical. By normalizing, and setting $x_1 = z-y$, we get the following equations :

$$\dot{x}_1 = u - x_2 \qquad |u| \le 1,$$

(18) $$\dot{x}_2 = \frac{v}{2p} \qquad |v| \le 1, \qquad p \text{ a parameter}$$

The playing space is

(19) $$|x_1| \le 1.$$

The real problem is to know whether we can, with v, insure (19) whatever u does. We formulate this as a game, with P trying to escape (violate (19)), E trying to forbid it. When escape occurs (the situation we shall consider), the payoff will be escape time :

$$J = \int_{t_o}^{t_1} dt .$$

The hamiltonian of this game is

$$H = 1 + \lambda_1(u-x_2) + \lambda_2 \frac{v}{2p} .$$

Away from singular surfaces, the optimal controls shall be

$$u = - \operatorname{sgn} \lambda_1, \quad v = \operatorname{sgn} \lambda_2,$$

so that optimal trajectories will be arcs of parabolas in the x space.

We readily find that the useable part of the capture set is made of two symetrical pieces :

$$x_1 = 1, \qquad x_2 < 1,$$

$$x_1 = -1, \qquad x_2 > -1.$$

From these points two pieces of barriers can be built. They are arcs of parabolas tangent to $|x_1| = 1$. To go further, one must consider the relative position of these parabolas.

We consider the case where p is smaller than 1, but not by much, say $0.8 \leq p < 1$. Then the two pieces of barrier are as shown on figure 3. (We use reverse axes for the benefit of space).

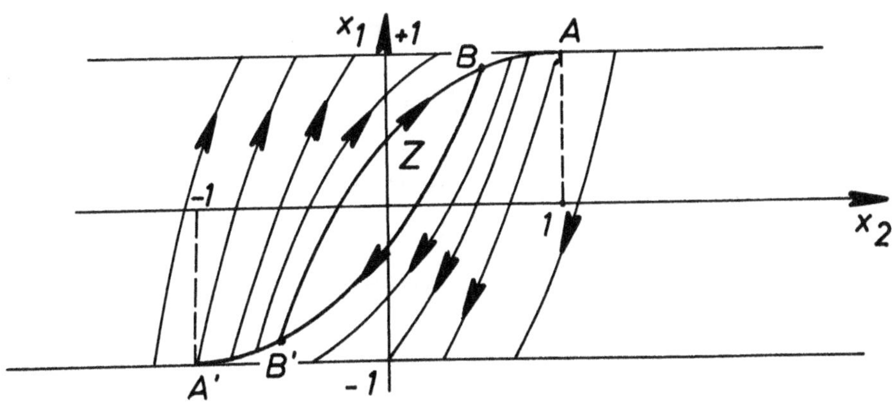

Fig.3. Barrier and primaries.

It seems that they define a closed region Z where the state is trapped, if E wants. However it is not so, because at points B and B' where the two pieces of barrier cut each other, E is not able to prevent the state to cross the two pieces at the same time. (In that respect, see [20] or [23] for a theory of junction of barriers). But before we completely solve the question thus raised, we must investigate more closely what happens outside of Z.

4.2. Lunge maneuver and equivocal line

The two pieces of barrier are the limits of two fields of primaries, one with $u = v = 1$, ends on $x_1 = 1$, one with $u = v = -1$ ends on $x_1 = -1$. Let B the point where the two pieces cut each other, in the region $x_1 > 0$, $x_2 > 0$. (The same things happen, by symmetry, on the other side).

The arc BA is a surface of discontinuity of the value function corresponding to these fields, continuous on the side of the field $(-1, -1)$, and thus with negative jump. We therefore look for an upper strategic pair $\varphi(x)$, $\psi(x, u)$ that avoids the jump.

The direction of the barrier and its normal (continuous side) are

$$\text{barrier}: \begin{pmatrix} 1 - x_2 \\ \frac{1}{2p} \end{pmatrix} \qquad \text{normal} \begin{pmatrix} -\frac{1}{2p} \\ 1-x_2 \end{pmatrix}$$

Thus,

$$\tilde{\mathcal{V}}_u(x) = \{v \mid -\tfrac{1}{2p}(u-x_2)+(1-x_2)\tfrac{v}{2p} \geq 0\} = \{v \geq \tfrac{u-x_2}{1-x_2}\}.$$

As, with the proposed value, in this region $\lambda_2 < 0$, it comes

$$\overline{\psi}(x, u) = \max\left(-1, \frac{u-x_2}{1-x_2}\right).$$

We place this in H and minimize in u. If we parametrize the field $(-1, -1)$ with $s = x_2(t_1)$, it finally comes

$$\overline{\varphi} = \overline{\psi} = +1, \quad \overline{H} = 2\,\frac{1+s-x_2}{1+s}, \quad \text{if} \quad 1 + s - x_2 \leq 0,$$

$$\overline{\varphi} = \overline{\psi} = -1, \quad \overline{H} = 0, \qquad \text{if} \quad 1 + s - x_2 \geq 0.$$

Therefore this field does not satisfy our sufficient conditions.

In the part of the arc BA where $1+s-x_2$ is negative, P can, by playing his barrier strategy, oblige E to do so, and let time to go decrease at a rate less than one in the process. The limit point C on the arc BA can be computed easily. At thid point, we have

$$0 = H(\overline{\varphi}, \overline{\psi}(\overline{\varphi})) \leq H(\varphi^*, \overline{\psi}(\varphi^*)) \leq H(\varphi^*, \psi^*) = 0 .$$

Hence, it can be the starting point of an equivocal line, with singularity in v. The theory gives us the equations of this line :

$$\dot{x}_1 = 1-x_2, \qquad \dot{\lambda}_1^- = \alpha(\lambda_1^+ - \lambda_1^-),$$

$$\dot{x}_2 = \frac{v}{2p}, \qquad \dot{\lambda}_2^- = \lambda_1^- + \alpha(\lambda_2^+ - \lambda_1^-).$$

(we have used superscipt + and − for outgoing and incoming trajectories) Equations (15) and (14) read

$$\lambda_2^- = 0 ,$$

$$(1-x_2)(\lambda_1^+-\lambda_1^-) + \frac{v}{2p}(\lambda_2^+ - \lambda_2^-) = 0 ,$$

We can also use the first integral $H(x, \lambda^-, \overline{\varphi}, \overline{\psi}) = 0$, which gives

$$\lambda_1^- = \frac{1}{x_2-1}.$$

The condition $\dot{\lambda}_2^- = 0$ would have given α, but we do not need it here .

We therefore have the differential equation for the equivocal line :

$$\dot{x}_2 = -\frac{1}{\lambda_2^+}[\lambda_1^+(1-x_2) +1].$$

It turns out that, in the field (−1, −1), λ_1^+ and λ_2^+ are easy to obtain as functions of the state, so that we can readily compute this line. It is a commutation line, with the incoming field having the same controls as the barrier (+1, +1), the outgoing field being (−1, −1).

As we integrate backward from C, the equivocal line cuts trajectories of the outgoing field with decreasing s. As the state approaches the barrier of that field, the magnitude of the gradient λ^+ tends to infinity. Therefore the equivocal line tends to become normal to it, and thus it reaches the barrier (−1, −1) tangentially at a point D.

The limit outgoing trajectory is the barrier DA'. However, with our definition of an open capture set, the barrier itself is not a capture trajectory. Therefore, from D, P should choose to keep $u = +1$ for awhile before switching.

4.3. The state constraint.

Consider a point on the barrier DA', close to D. If capture should occur by first following the barrier, then the optimal strategy would be to switch to (+1, +1) at C', symmetric point to C, and then follow the new parabola until capture. This takes much more time than the optimal strategies we have proposed from D. Therefore P would rather try to reach D. Now, by playing $u = +1$ he can actually insure that $\dot{x}_1$ be positive.

If the state must actually reach D before leaving the zone Z, along the equivocal line, then the barrier is a surface of discontinuity of the value, discontinuous on the outside of Z as we have seen, with a negative jump. Therefore, E must prevent the state from crossing the barrier, which in that region will act as a state constraint.

We can apply the previous theory, with

$$\tilde{\mathcal{V}}_u(x) = \{v| \leq \frac{u-x_2}{1+x_2}\}$$

and, as in that region, by continuity with the incoming field of the equivocal line, λ_2 is positive, we find :

$$\bar{\psi}(x, u) = \min\left(1, \frac{u-x_2}{1+x_2}\right)$$

Starting from P, we can easily integrate a trajectory that traverses the barrier in the opposite direction from the "natural" one. The theory shows that we must still have, on the barrier

$$\lambda_2 = 0$$

and again, the first integral $H=0$ gives λ_1 as previously.

The state constraint is joined by the same incoming field (+1, +1) as the equivocal line.

This strategy is optimal, along the barrier, as long as it yields

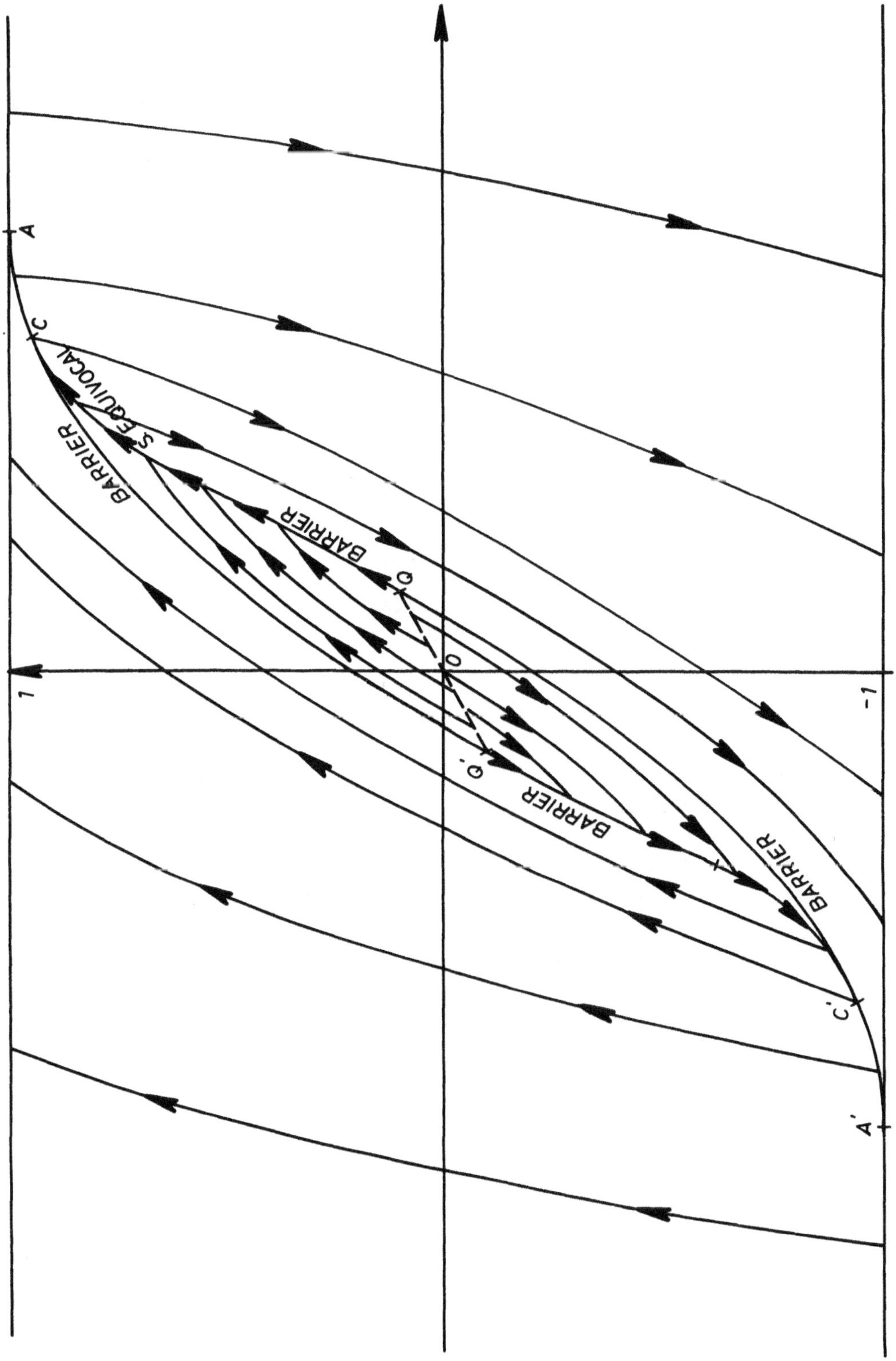

Fig. 4

a shorter capture time than following the barrier in its natural direction, and switching at C'. This defines a point Q where from the two strategies yield the same time to go. For the range of parameter values we have set here, Q is the starting point of a simple dispersal line separating the field (+1, +1) from the field (-1, -1), and readily computed using the requirement that it be normal to $(\lambda^+-\lambda^-)$.

For smaller values of p the situation is more complicated, Q being the starting point of a new equivocal line, itself followed smoothly by a dispersion line. Then, this extremely simple game may have a seven-stage optimal capture trajectory.

BIBLIOGRAPHY

[1] R. ISAACS "Differential Games" Rand Reports, 1954, 1965, and Differential Games, Wiley, N.Y., 1965.

[2] J.V. BREAKWELL "Some differential games with interesting discontinuities", Stanford University, Department of Aeronautics & Astronautics. 1973.

[3] W.H. FLEMING "The convergence problem for differential games", Jal of Math. Analysis and Applications, vol 3., pp 102-116,1961.

[4] W.H. FLEMING "The Convergence problem for differential games, II" Advances in Game Theory, Princeton University press, pp 195-210, 1964.

[5] P. VARAIYA "The existence of solutions to a differential game" SIAM Jal on Control Vol 5, pp 153-162, 1967.

[6] E. ROXIN "On Varaiya's definition of a differential game" US-Japan seminar on differential and functional equations, Mineapolis, 1967.

[7] P. VARAIYA & J. LIN "Existence of saddle points in differential games" SIAM Jal on Control Vol 7, pp 141-157, 1969.

[8] A. FRIEDMAN : Differential games. Wiley, New York, 1971.

[9] N.S. PONTRYAGIN "Linear differential games" Soviet Math. Doklady vol 8, pp 769-771 and pp 910, 912, 1967.

[10] N N KRASOVSKYI. Positional differential games. English translation to appear.

[11] J.F. MASLE "Problèmes qualitatifs et quantitatifs liés en jeux differentiels", Thesis, Université de Paris IX, 1976.

[12] W.T. BOARDMAN III " An analytical investigation on the state constraints in Isaacs Isotropic Rocket Game. Stanford University, Engineer thesis, 1968.

[13] J.V. BREAKWELL & A.W. MERZ "Toward a complete solution of the Homicidial Chauffeur game" First International Conference on the Theory and Applications of Differential Games, Amherst 1969.

[14] J.V. BREAKWELL "Complete solution of the Dolichobrachistochrone problem", unpublished, lectures notes, Stanford University, 1968.

[15] L. BERKOVITZ "Lectures on Differential games" in Differential games and related topics, Kuhn & Szego ed., North Holland, 1971.

[16] A. BLAQUIERE et al. Qualitative and Quantitative games, Academic Press, 1970.

[17] A.W. MERZ "The homicidial chauffeur - a differential game", PhD Thesis, Stanford University, 1971.

[18] W.E SCHMITENDORF "Differential Games without pure strategy saddle point solutions", Jal of Differential Equations, pp 81-92, 1976.

[19] S.S.L. CHANG, T.K.C PENG : "Adaptive guaranteed cost control of systems with uncertain parameters". IEEE Trans. AC 17, pp 474-482, 1972.

[20] P. BERNHARD "New results about corners in differential games, including state constraints". 6th IFAC world Congress, Cambridge Mass. 1975.

[21] P. COLLETER. Thèse d'Ingénieur docteur, University Paris IX 1977. (to appear.

[22] P. BERNHARD. "Commande Optimale, décentralisation et jeux dynamiques"

[23] P. BERNHARD "Corner Conditions for differential games". 5th IFAC world congress, Paris, 1972.

DIFFERENTIAL GAMES WITH PIECE-WISE CONTINUOUS TRAJECTORIES

A. Blaquière
Université de Paris 7

PART I : A PROBLEM OF IMPULSIVE OPTIMAL CONTROL

1.1. PROBLEM STATEMENT

1.1.1. Strategies and Paths

We shall be concerned with a *dynamical system* defined by its *state*, a set of n real numbers, $x = (x_1, \ldots x_n) \in R^n$, which change in a prescribed manner with the passing of time t. We shall suppose that the state lies in some open subset X of R^n, and that one of its components, say x_n, is time t. The evolution of the state is influenced, or as we say controlled, by an agent J_o through his choice of a *strategy* s in a prescribed *strategy set* S_o. First, we shall define this strategy set and the rule that governs the motion of the state.

Let U, M be prescribed non-empty open subsets of R^{d_1} and R^{d_2}, respectively. Let K_u and K_μ be prescribed non-empty subsets of U and M, respectively. Let P and Π be prescribed sets of functions defined on X with range in K_u and K_μ, respectively. Let Δ be the collection of all closed† subsets of X.

Definition 1.1. The strategy set of J_o is $S_o = \Delta \times P \times \Pi$.

In other words, J_o will influence the evolution of the state through his choice of

(a) a closed subset of X, say $Y \in \Delta$; and

(b) a pair of functions defined on X, say $(p(\cdot), \pi(\cdot)) \in P \times \Pi$.

Now let $f(\cdot)$ and $g(\cdot)$ be prescribed functions of class C^1 on $X \times U$ and $X \times M$, respectively; namely

$$f(\cdot) : \begin{cases} X \times U \to R^n \\ (x,u) \mapsto f(x,u) \end{cases}$$

$$g(\cdot) : \begin{cases} X \times M \to R^n \\ (x,\mu) \mapsto g(x,\mu) \end{cases}$$

with $f = (f_1, \ldots f_n)$, $g = (g_1, \ldots g_n)$; and $f_n(x,u) \equiv 1$ and $g_n(x,\mu) \equiv 0$.

† in the topology induced by R^n on X.

<u>Definition 1.2.</u> A strategy s, $s = (Y, p(\cdot), \pi(\cdot)) \in S_o$ is *admissible* if and only if

$$x \in Y \Rightarrow x + g(x, \pi(x)) \in X - Y$$

Let S be the set of all admissible strategies.

<u>Definition 1.3.</u> A function $x(\cdot): [t_o, t_1] \to \overline{X}$†, defined on some interval of time $[t_o, t_1]$ is a *path in* R^n *generated by* $s = (Y, p(\cdot), \pi(\cdot)) \in S$ *from initial state* $x^o \in X$ if and only if

(i) $x(t_o) = x^o$; and

(ii) $x(\cdot)$ is piece-wise continuous on $[t_o, t_1]$ - Let $T[t_o, t_1]$ denote the set of its discontinuity points - and

(iii) $x(t) = x(t-0)$ for $t \neq t_o$; and

(iv) $t \in T[t_o, t_1] \Rightarrow x(t) \in Y$ and

$$x(t+0) = x(t) + g(x(t), \pi(x(t)));$$ and

(v) for all t which are not discontinuity points, except possibly at $t = t_1$

$x(t) \in X - Y$; and

(vi) $x(\cdot)$ is differentiable, and

$$\frac{dx(t)}{dt} = f(x(t), p(x(t)))$$

for all $t \in [t_o, t_1]$, except on a subset of $[t_o, t_1]$ at most denumerable.

It follows from (iii) that $t_1 \notin T[t_o, t_1]$.

<u>1.1.2. Playability</u>

We shall suppose that J_o desires to steer the state from a given initial state x^o to a state belonging to a prescribed *target* set, $\theta \subset \partial X$.

<u>Definition 1.4.</u> A path $x(\cdot): [t_o, t_1] \to \overline{X}$ generated by $s \in S$ from initial state $x^o \in X$ is a *terminating path* if and only if $x(t_1) \in \theta$.

---

† $\overline{X}$ denotes the closure of X in the topology of R^n

Definition 1.5. A strategy $s \in S_o$ is *playable at* x^o, $x^o \in X$, if and only if it is admissible and it generates a terminating path from x^o.

Let $J(x^o)$ denote the set of all strategies playable at x^o, and $I(x^o,s)$ the set of all terminating paths generated from x^o by a playable strategy $s \in J(x^o)$.

1.1.3. Cost of a Path

Let $f_o(\cdot)$ and $g_o(\cdot)$ be prescribed functions of class C^1 on $X \times U$ and $X \times M$, respectively; namely

$$f_o(\cdot) : \begin{cases} X \times U \to R \\ (x,u) \mapsto f_o(x,u) \end{cases}$$

$$g_o(\cdot) : \begin{cases} X \times M \to R \\ (x,\mu) \mapsto g_o(x,\mu) \end{cases}$$

Later on it will be convenient to let $F(\cdot) = (f_o(\cdot), f(\cdot))$ and $G(\cdot) = (g_o(\cdot), g(\cdot))$.

Let $\tilde{\theta}_o(\cdot)$ be a prescribed function of class C^1 on some domain $D \subset R^n$ containing $\overline{X}$

$$\tilde{\theta}_o(\cdot) : \begin{cases} D \to R \\ x \mapsto \tilde{\theta}_o(x) \end{cases}$$

Let $\theta_o(\cdot)$ coincide with $\tilde{\theta}_o(\cdot)$ on θ, and $\theta_o(x) = 0$ for all $x \in X$.

The cost of a path $x(\cdot): [t_o,t_1] \to \overline{X}$ generated by $s = (Y, p(\cdot), \pi(\cdot)) \in S$ from initial state $x^o \in X$ is

$$V(x^o,s,x(\cdot)) = \theta_o(x(t_1)) + \int_{t_o}^{t_1} f_o(x(t), p(x(t)))dt$$

$$+ \sum_{t \in T[t_o,t_1]} g_o(x(t), \pi(x(t)))$$

1.1.4. Optimality of a Strategy

Let us suppose that J_o desires to minimize the cost of transfer from an initial state to the target.

Definition 1.6. A strategy s^* is *optimal on* X if and only if

(i) s^* is playable al all $x^o \in X$; and

(ii) there exists a function $V^*(\cdot)$: $X \to R$ such that for all $x^o \in X$ and all $x(\cdot) \in I(x^o, s^*)$

$V(x^o, s^*, x(\cdot)) = V^*(x^o)$, and

(iii) for all $x^o \in X$, and all $s \in J(x^o)$, and all $x(\cdot) \in I(x^o,s)$

$V^*(x^o) \leqslant V(x^o, s, x(\cdot))$

By letting $V^*(x^o) = \theta_o(x^o)$ for all $x^o \in \theta$, $V^*(\cdot)$ is defined on $X \cup \theta$.

1.1.5. Limiting Surfaces and Paths in Augmented State Space.

Let $y = (x_o, x) = (x_o, x_1, \ldots x_n) \in R^{n+1}$ denote an augmented state. Since function $V^*(\cdot)$ is defined on $X \cup \theta$ we can state

Definition 1.7. A *limiting surface* is

$$\Sigma(C) = \{ y: \quad x \in X \cup \theta, \quad x_o + V^*(x) = C \}$$

where C is a constant parameter.

A given limiting surface $\Sigma(C)$ separates $R \times (X \cup \theta)$ into two disjoint sets

$$A/\Sigma(C) = \{ y: \quad x \in X \cup \theta, \quad x_o + V^*(x) > C \}$$
$$B/\Sigma(C) = \{ y: \quad x \in X \cup \theta, \quad x_o + V^*(x) < C \}$$

Definition 1.8. A function $y(\cdot) = (x_o(\cdot), x(\cdot))$: $[t_o, t_1] \to R \times \overline{X}$ is a *path in* R^{n+1} *generated by* $s = (Y, p(\cdot), \pi(\cdot)) \in S$ *from initial state* $y^o = (x_o^o, x^o) \in R \times X$ if and only if $x(\cdot)$: $[t_o, t_1] \to \overline{X}$ is a path in R^n generated by s from x^o and, for all $t \in [t_o, t_1]$,

$$x_o(t) + \theta_o(x(t_1)) + \int_t^{t_1} f_o(x(\tilde{t}), p(x(\tilde{t})))d\tilde{t}$$

$$+ \sum_{\tilde{t} \in T[t, t_1]} g_o(x(\tilde{t}), \pi(x(\tilde{t}))) = C$$

where again, C is a constant parameter, and $T[t, t_1] = [t, t_1] \cap T[t_o, t_1]$.

By varying C one generates a family of limiting surfaces and a family of paths in R^{n+1} .

From the definition of a path in R^n it follows that

(i) $y(\cdot)$ is piece-wise continuous on $[t_o, t_1]$, and $T[t_o, t_1]$ is the set of its discontinuity points; and

(ii) for all $t \in T[t_o, t_1]$, $t \neq t_o$

$$y(t) = y(t-0) = \lim_{\substack{\varepsilon \to 0 \\ \varepsilon < 0}} y(t+\varepsilon), \text{ and}$$

(iii) for all $t \in T[t_o, t_1]$

$$y(t+0) = \lim_{\substack{\varepsilon \to 0 \\ \varepsilon > 0}} y(t+\varepsilon) = y(t) + G(x(t), \pi(x(t))), \text{ and}$$

(iv) $y(\cdot)$ is differentiable, and

$$\frac{dy(t)}{dt} = F(x(t), p(x(t)))$$

for all $t \in [t_o, t_1]$ except on a subset of $[t_o, t_1]$ at most denumerable.

From now on we shall place certain restrictions on the sets P and Π. They arise from requirements inherent to the subsequent mathematical development.

Assumption 1.1. $(\bar{p}(\cdot), \bar{\pi}(\cdot))$, $(\bar{\bar{p}}(\cdot), \bar{\bar{\pi}}(\cdot)) \in P \times \Pi$, $\bar{x} \in X$ and $(p(\cdot), \pi(\cdot))$ such that

$$p(x) = \bar{p}(x),\ \pi(x) = \bar{\pi}(x) \quad \text{for} \quad x_n < \bar{x}_n$$

$$p(x) = \bar{\bar{p}}(x),\ \pi(x) = \bar{\bar{\pi}}(x) \quad \text{for} \quad x_n \geqslant \bar{x}_n$$

implies that

$$(p(\cdot), \pi(\cdot)) \in P \times \Pi \ .$$

We shall leave it to the reader to verify that Assumption 1.1. assures that paths can be "joined". Namely, consider paths $\bar{x}(\cdot): [t_o, t_1] \to X \cup \theta$ and $\bar{\bar{x}}(\cdot): [t_1, t_2] \to X \cup \theta$, generated by admissible strategies $\bar{s}$ and $\bar{\bar{s}}$, respectively, such that $\bar{x}(t_1) = \bar{\bar{x}}(t_1)$. Then there exists an admissible strategy s generating a path $x(\cdot): [t_o, t_2] \to X \cup \theta$, such that $x(t) = \bar{x}(t)$ for all $t \in [t_o, t_1]$ and $x(t) = \bar{\bar{x}}(t)$ for all $t \in [t_1, t_2]$. Indeed, a similar property holds for paths in R^{n+1}.

The control problem we are considering in this paper is a spacial case of a game problem, when the number of players is one. When the number of players is arbitrary, and for a wide class of problems, we have defined in earlier publications [see for instance (1) (2)] a family of *game surfaces*. A limiting surface is a game surface in the special case of one-player games. General properties of game surfaces and paths have been stated in (1) (2) (3) . As our problem, now, is a special case of the one considered in (1) (2) (3), we shall state here without proof a basic theorem.

Theorem 1.1. No point of a path in R^{n+1} *emanating from an initial point* $y^o \in \Sigma(C)$ *is in* $B/\Sigma(C)$; *and all points of a path in* R^{n+1} *generated by a strategy* s^* *optimal on* X, *emanating from an initial point* $y^o \in \Sigma(C)$, *are in* $\Sigma(C)$.

1.2. NECESSARY CONDITIONS FOR AN OPTIMAL STRATEGY

1.2.1. Assumptions

Before proceeding to a derivation of necessary conditions for a strategy to be optimal, we need

Assumption 1.2. There exists a strategy s^* optimal on X, namely $s^* = (Y^*, p^*(\cdot), \pi^*(\cdot))$.

Let $x^*(\cdot): [t_i, t_j] \to \bar{X}$ be a path generated by s^*, such that $T[t_i, t_j] = t_c$; that is, a path along which there is a single discontinuity point t_c, $t_i \leq t_c < t_j$. Two cases need be considered, namely

Case (a): $t_c \neq t_i$ and accordingly, from the definition of a path,

$$x^*(t_c) \in \partial Y^*; \text{ and}$$

Case (b): $t_c = t_i$

Assumption 1.3. There exists an open ball $B(x^*(t_c+0)) \subset X - Y^*$ with center $x^*(t_c+0)$, on which $p^*(\cdot)$ and $V^*(\cdot)$ are of class C^1.

Assumption 1.4. There exists an open ball $B(x^*(t_c)) \subset X$ with center $x^*(t_c)$ such that

in Case (a) : $t_c \neq t_i$,

(i) $p^*(\cdot)$ agrees on $(X - Y^*) \cap B(x^*(t_c))$ with a function - say $\bar{p}(\cdot)$ - which is of class C^1 on $B(x^*(t_c))$, and

(ii) $V^*(\cdot)$ is C^1 on $(X - Y^*) \cap B(x^*(t_c))$; and *in Case (b)* : $t_c = t_i$, $V^*(\cdot)$ is C^1 on $B(x^*(t_c))$, except if there exists a path $x^{**}(\cdot)\colon [t_i - \varepsilon, t_j] \to \overline{X}$, $\varepsilon > 0$, generated by s^*, which coincides with $x^*(\cdot)$ on $[t_i, t_j]$, in which case $x^*(t_c) \in \partial Y^*$ and we keep the same assumption as for Case (a).

Assumption 1.5. (i) The set of all constant functions on X with range in K_μ is a subset of Π, and (ii) the set of all constant functions on X with range K_u is a subset of P.

A set Ω in R^{n+1} is directionally convex with respect to vector w^o, or equivalently it is x_o^+ - *directionally convex*, if for all $y^1, y^2 \in \Omega$ and all $\nu \in [0,1]$ there exists a χ, $-\infty < \chi \leqslant 0$, such that

$$y^1 + \nu(y^2 - y^1) + \chi w^o \in \Omega$$

where $w^o = (1,0,\ldots 0)$ is a unit vector in the x_o - direction.
Ω is x_o^- - *directionally convex* if it is directionally convex with respect to $-w^o = (-1, 0, \ldots 0)$.

Assumption 1.6. The set $\Omega(y^*(t_c))$ is x_o^+ - directionally convex,

$$\Omega(y^*(t_c)) = \{\, y^*(t_c) + G(x^*(t_c), \mu)\colon \ \mu \in K_\mu \,\}$$

where $y^*(\cdot) = (x_o^*(\cdot), x^*(\cdot))$ is a path in R^{n+1} generated by s^*, whose projection in R^n is $x^*(\cdot)$.

1.2.2. Theorem 1.2.

Here we shall state necessary conditions for a strategy to be optimal. Let $\lambda(\cdot)\colon [t_i, t_j] \to R^{n+1}$ be a piece-wise continuous function, with $\lambda(t) = \lambda(t-0)$, for all $t \in (t_i, t_j]$, and let

$$H(\lambda, x, u) = \sum_{\alpha=0}^{n-1} \lambda_\alpha f_\alpha(x,u)\,, \qquad \lambda = \lambda(t)$$

$$H_c(x, \mu) = \sum_{\alpha=0}^{n-1} \lambda_\alpha(t_c + 0)\, g_\alpha(x, \mu)$$

Theorem 1.2. *If Assumptions 1.1 - 1.6 are satisfied and* $t_c \neq t_i$, *there exists* $\tau > 0$ *and* $\lambda_\alpha(\cdot)$, $\alpha = 0,1,\ldots n-1$, *solutions of (1.1) on* $[t_c - \tau, t_c)$ *and* $(t_c, t_c + \tau]$,

$$\dot{\lambda}_\alpha = - \left.\frac{\partial H(\lambda,x,u)}{\partial x_\alpha}\right|_{x = x^*(t),\ u = p^*(x)} \tag{1.1}$$

with

$$\lambda_\alpha(t_c) = \lambda_\alpha(t_c + 0) + \left.\frac{\partial H_c(x,\mu)}{\partial x_\alpha}\right|_{x = x^*(t_c),\ \mu = \pi^*(x)} \tag{1.2}$$

such that

(i) $\min\limits_{u \in K_u} H(\lambda(t), x^*(t), u) = H(\lambda(t), x^*(t), p^*(x^*(t)))$

for all $t \in [t_c - \tau, t_c + \tau]$, $t \neq t_c$; *and*

(ii) $\min\limits_{\mu \in K_\mu} H_c(x^*(t_c), \mu) = H_c(x^*(t_c), \pi^*(x^*(t_c)))$; *and*

(iii) $\min\limits_{u \in K_u} H(\lambda(t_c + 0), x^*(t_c + 0), u) - \min\limits_{u \in K_u} H(\lambda(t_c), x^*(t_c), u) =$

$$= \left.\frac{\partial H_c(x,\mu)}{\partial x_n}\right|_{x = x^*(t_c),\ \mu = \pi^*(x)}$$

and

(iv) $\lambda_o(t) = 1$ *for all* $t \in [t_c - \tau, t_c + \tau]$.

If $t_c = t_i$ *and there does not exist a path* $x^{**}(\cdot): [t_i - \varepsilon, t_j] \to \bar{X}$, $\varepsilon > 0$, *generated by* s^*, *which coincides with* $x^*(\cdot)$ *on* $[t_i, t_j]$, *then* (i) *above is satisfied for all* $t \in (t_c, t_c + \tau]$, *and* (ii) *is satisfied, and* (iv) *is satisfied for all* $t \in [t_c, t_c + \tau]$, *and*

(v) $\min\limits_{u \in K_u} H(\lambda(t_c+0), x^*(t_c+0), u) - \min\limits_{u \in K_u} H(\lambda(t_c), x^*(t_c), u) \leqslant$

$$\leqslant \left.\frac{\partial H_c(x,\mu)}{\partial x_n}\right|_{x = x^*(t_c),\ \mu = \pi^*(x)}$$

Indeed, if $t_c = t_i$, and path $x^{**}(\cdot)$ exists, then $x^*(\cdot)$ is replaced by $x^{**}(\cdot)$ and conditions (i) - (iv) of Theorem 1.2. are applied to that new path, with t_i replaced by $t_i - \varepsilon$.

As in the case of optimal control problems with continuous trajectories, Theorem 1.2. needs be complemented by a *transversality condition* which must hold at the point where an optimal path in R^n reaches θ. The arguments for deriving the transversality condition in our problem are the same as in the case of continuous trajectories; the assumptions they require concerning θ, and the condition itself, are the same. Here we shall simplify the transversality condition by assuming that θ is a domain in the (n-1) - dimensional plane $x_n = T$. Then the transversality condition reads

$$\lambda_\alpha(T) = \partial\theta_o(x)/\partial x_\alpha \Big|_{x=x^*(T)}, \qquad \alpha = 1,\dots n-1$$

<u>Proof.</u> Consider path $x^*(\cdot): [t_i,t_j] \to \bar{X}$ generated by s^*, such that $T[t_i,t_j] = t_c$, $t_c \neq t_i$, and suppose that the interval $[t_i,t_j]$ is sufficiently small so that

$$x^*(t) \in B(x^*(t_c)) \cup B(x^*(t_c+0)) \quad \text{for all} \quad t \in [t_i,t_j]$$

where $B(x^*(t_c+0)$ and $B(x^*(t_c))$ satisfy Assumptions 1.3 and 1.4, respectively.

Consider also a path $y^*(\cdot) = (x_o^*(\cdot),\ x^*(\cdot))$ in R^{n+1}, generated by s^*, whose projection in R^n is $x^*(\cdot)$. From the definition of a limiting surface and Theorem 1.1, there is a limiting surface $\Sigma(C)$ such that

$$y^*(t) \in \Sigma(C) \qquad \text{for all} \quad t \in [t_i,t_j] \tag{1.3}$$

Then consider a time t_k, $t_i \leq t_k < t_c$, and a solution $y^k(\cdot): [t_k,t_k+\alpha] \to R^{n+1}$ of†

$$\dot{y} = F(y,u), \qquad u \in K_u$$

such that $y^k(t_k) = y^*(t_k) = y^k$. For sufficiently small α this solution exists and is unique, as y^k is an interior point of $R \times X$ and $F(\cdot)$ is C^1 on $X \times U$. It follows at once from Theorem 1.1. and (ii) of Assumption 1.5, that

$$y^k(t) \in (A/\Sigma(C) \cup \Sigma(C) \qquad \text{for all} \quad t \in [t_k,\ t_k+\alpha] \tag{1.4}$$

for sufficiently small α .

Let $\Phi(\cdot): R \times (X \cup \theta) \to R$ be the function such that $\Phi(y) = x_o + V^*(x)$ for all $y = (x_o,x) \in R \times (X \cup \theta)$. By definition of $\Sigma(C)$ and $A/\Sigma(C)$ we have

$$\Phi(y) = C \iff y \in \Sigma(C)$$
$$\Phi(y) \geqslant C \iff y \in (A/\Sigma(C)) \cup \Sigma(C)$$

† Here we let $F(y,u) = F(x,u)$

As a consequence of (ii) of Assumption 1.4, Case (a), there exists an open ball $B(x^*(t_k)) \subset X$, with center $x^*(t_k)$, such that function $\Phi(\cdot)$ is of class C' on $R \times B(x^*(t_k))$. Consequently, for sufficiently small α, relations (1.4) and (1.3) result in

$$\operatorname{grad} \Phi(y^k) \cdot F(y^k,u) \geqslant 0 \qquad \text{for all} \quad u \in K_u \tag{1.5}$$

$$\operatorname{grad} \Phi(y^k) \cdot F(y^k, p^*(x^k)) = 0 \tag{1.6}$$

where

$$\operatorname{grad} \Phi(y) = \left(1, \frac{\partial V^*(x)}{\partial x_1}, \ldots \frac{\partial V^*(x)}{\partial x_n}\right)$$

Let $\lambda(t) = (\lambda_0(t), \lambda_1(t), \ldots \lambda_n(t)) = \operatorname{grad} \Phi(y^*(t))$, $t_i \leqslant t < t_c$. In view of (i) of Assumption 1.4, Case (a), and of the fact that the constraints on u do not depend on the state, one can prove easily by similar arguments as in [1], for regular optimal paths, that $\lambda_\alpha(\cdot)$, $\alpha = 0,1,\ldots n$, on $[t_i, t_c)$ is a solution of adjoint equations

$$\dot{\lambda}_\alpha = - \sum_{\beta=0}^{n} \left.\frac{\partial f_\beta(x,u)}{\partial x_\alpha} \lambda_\beta \right|_{x = x^*(t),\ u = p^*(x)}$$

from which we deduce (1.1) since $f_n(x,u) \equiv 1$.

As conditions (1.5) and (1.6) hold for all $t_k \in [t_i, t_c)$, by dropping index k, they can be rewritten

$$\lambda_n(t) + \sum_{\alpha=0}^{n-1} \lambda_\alpha(t) f_\alpha(x^*(t),u) \geqslant 0 \qquad \text{for all} \quad u \in K_u \tag{1.7}$$

$$\lambda_n(t) + \sum_{\alpha=0}^{n-1} \lambda_\alpha(t) f_\alpha(x^*(t), p^*(x^*(t))) = 0 \tag{1.8}$$

for all $t \in [t_i, t_c)$. Another way of writing (1.7) and (1.8) is

$$\lambda_n(t) + \min_{u \in K_u} H(\lambda(t), x^*(t), u) = \lambda_n(t) + H(\lambda(t), x^*(t), p^*(x^*(t))) = 0 \tag{1.9}$$

for all $t \in [t_i, t_c)$, from which condition (i) of Theorem 1.2 is deduced for $t \in [t_i, t_c)$. Similar arguments for $t_c < t_k \leqslant t_j$ conclude the proof of the first part of Theorem 1.2 .

Now, as $V^*(\cdot)$ is C^1 on $B(x^*(t_c + 0))$ according to Assumption 1.3, $\lambda(t_c + 0)$ is defined; that is

$$\lambda(t_c+0) = \text{grad } \Phi(y^*(t_c+0))$$

Because of Theorem 1.1. and (i) of Assumption 1.5. we have

$$\Omega(y^*(t_c)) \cap (B/\Sigma(C)) = \phi \qquad (1.10)$$

Because of Assumption 1.6, for all $\mu \in K_\mu$, and for all $\varepsilon \in [0,1]$, there exists a χ, $-\infty < \chi \leqslant 0$, such that

$$y^*(t_c+0) + \varepsilon\xi(\mu) + \chi\, w^\circ \in \Omega(y^*(t_c))$$

where $\xi(\mu) = G(x^*(t_c),\mu) - G(x^*(t_c),\pi^*(x^*(t_c)))$, and, because of (1.10),

$$y^*(t_c+0) + \varepsilon\xi(\mu) + \chi\, w^\circ \notin B/\Sigma(C) \qquad (1.11)$$

As $y^*(t_c+0)$ is an interior point of $R\times X$, it follows from (1.11) that, for ε sufficiently small,

$$y^*(t_c+0) + \varepsilon\xi(\mu) \in (A/\Sigma(C)) \cup \Sigma(C)$$

and accordingly

$$\Phi(y^*(t_c+0)) + \varepsilon\xi(\mu)) \geqslant C$$

from which we deduce

$$\Phi(y^*(t_c+0) + \varepsilon\xi(\mu)\cdot\text{grad } \Phi(y^*(t_c+0)) + o(\varepsilon) \geqslant C$$

where

$$\lim_{\varepsilon\to 0} \frac{|o(\varepsilon)|}{\varepsilon} = 0$$

Since $y^*(t_c+0) \in \Sigma(C)$ and accordingly $\Phi(y^*(t_c+0)) = C$, we obtain, upon division by ε, and by letting ε tend to zero

$$\xi(\mu)\cdot\text{grad } \Phi(y^*(t_c+0)) \geqslant 0 \qquad (1.12)$$

(1.12) rewrites

$$(G(x^*(t_c),\mu) - G(x^*(t_c),\pi^*(x^*(t_c)))) \cdot \lambda(t_c+0) \geqslant 0$$

from which we deduce

$$G(x^*(t_c),\mu) \cdot \lambda(t_c+0) \geqslant G(x^*(t_c),\pi^*(x^*(t_c))) \cdot \lambda(t_c+0)$$

Since $g_n(x,\mu) \equiv 0$, we obtain at last

$$\sum_{\alpha=0}^{n-1} \lambda_\alpha(t_c+0)g_\alpha(x^*(t_c),\mu) \geqslant \sum_{\alpha=0}^{n-1} \lambda_\alpha(t_c+0)g_\alpha(x^*(t_c),\pi^*(x^*(t_c)))$$

for all $\mu \in K_\mu$, which is equivalent to condition (ii) of Theorem 1.2.

In order to prove condition (iii) of Theorem 1.2, let us consider a point y^k at time t_k in a neighborhood in $\Sigma(C)$ of point $y^*(t_i)$. Let

$$y^k = y^*(t_i) + \varepsilon\eta^i + o(\varepsilon) \in \Sigma(C) \tag{1.13}$$

where $y^*(t_i)+\eta^i$ lies in the tangent plane of $\Sigma(C)$ at $y^*(t_i)$. This tangent plane exists since $\operatorname{grad}\Phi(y^*(t_i))$ is defined according to (ii) of Assumption 1.4, Case (a). Later on we shall let ε tend to zero and

$$\lim_{\varepsilon\to 0} \frac{\|o(\varepsilon)\|}{\varepsilon} = 0$$

Then consider the solution $y^k(\cdot): [t_k, t_k+\tau] \to R^{n+1}$, $\tau = t_c - t_i$, of

$$\dot{y} = F(y, \bar{p}(y))$$

where $\bar{p}(y) = \bar{p}(x)$ is defined in Assumption 1.4, such that $y^k(t_k) = y^k$. For sufficiently small ε this solution exists and is unique according to Assumption 1.4. Let $y^- = y^k(t_k+\tau)$.

From the dependence on initial conditions of the solution $y^k(\cdot)$ it follows that

$$y^- = y^*(t_c) + \varepsilon\eta^- + o(\varepsilon), \qquad \eta^- = (\eta_o^-, \eta_1^-, \dots \eta_n^-)$$

where $\lim_{\varepsilon\to 0} \frac{\|o(\varepsilon)\|}{\varepsilon} = 0$ and $\eta^- = A(t_i,t_c)\eta^i$ where $A(t_i,t_c)$ is a linear non singular transformation.

Because of Assumption 1.4

$$\lambda(t_c) = \lim_{\substack{t\to t_c\\ t<t_c}} \operatorname{grad}\Phi(y^*(t))$$

is defined, and

$$\operatorname{grad}\Phi(y^*(t_i))\cdot\eta^i = 0 \Rightarrow \lambda(t_c)\cdot\eta^- = 0 \tag{1.14}$$

Let T^i denote the n-dimensional tangent plane of $\Sigma(C)$ at $y^*(t_i)$. Since $A(t_i,t_c)$ is a linear non singular transformation, the set

$$T^c = \{y^*(t_c) + \eta^- : \ \eta^- = A(t_i,t_c)\eta^i,\ y^*(t_i) + \eta^i \in T^i\}$$

is an n-dimensional plane, and it follows from (1.14) that $\lambda(t_c)$ is perpendicular to T^c.

Now let $\mu^c = \pi^*(x^*(t_c))$, $y^- = (x_c^-, x^-)$, and

$$y^+ = y^- + G(x^-, \mu^c)$$

Since $G(\cdot)$ is C^1 on $X \times M$

$$y^+ = y^*(t_c + 0) + \varepsilon\eta^+ + o(\varepsilon), \qquad \eta^+ = (\eta_o^+, \eta_1^+, \ldots \eta_n^+)$$

where $\lim_{\varepsilon \to 0} \frac{\| o(\varepsilon) \|}{\varepsilon} = 0$ and

$$\eta_\alpha^+ = \eta_\alpha^- + \sum_{\beta=0}^{n} \left. \frac{\partial g_\alpha(x, \mu^c)}{\partial x_\beta} \right|_{x = x^*(t_c)} \eta_\beta^- \tag{1.15}$$

$\alpha = 0, 1, \ldots n$.

In other words $\eta^+ = M^c \eta^-$ where M^c is a linear transformation defined by (1.15), which needs not be non-singular.

Let us define $\lambda^- = (\lambda_o^-, \lambda_1^-, \ldots \lambda_n^-)$ by

$$\lambda_\alpha^- = \lambda_\alpha(t_c + 0) + \sum_{\beta=0}^{n} \left. \frac{\partial g_\beta(x, \mu^c)}{\partial x_\alpha} \right|_{x = x^*(t_c)} \lambda_\beta(t_c + 0) \tag{1.16}$$

$\alpha = 0, 1, \ldots n$.

One can readily verify that

$$\sum_{\alpha=0}^{n} \lambda_\alpha^- \eta_\alpha^- = \sum_{\alpha=0}^{n} \lambda_\alpha(t_c + 0)\, \eta_\alpha^+ \tag{1.17}$$

and that $\lambda_o^- = \lambda_o(t_c + 0) = 1$.

Now let us rely upon Theorem 1.1 again. As $y^*(t_i)$, $y^*(t_c)$ and $y^*(t_c + 0)$ are interior points of $R \times X$, for sufficiently small $|\varepsilon|$ we have

$$y^k \in \Sigma(C) \Rightarrow y^- \in (A/\Sigma(C)) \cup \Sigma(C) \Rightarrow y^+ \in (A/\Sigma(C)) \cup \Sigma(C)$$

from which it follows that

$$\Phi(y^+) = \Phi(y^*(t_c + 0)) + \varepsilon\eta^+ \cdot \text{grad}\, \Phi(y^*(t_c + 0)) + o(\varepsilon) \geqslant C$$

where $\lim_{\varepsilon\to 0} \frac{\|o(\varepsilon)\|}{\varepsilon} = 0$, and since $\Phi(y^*(t_c+0)) = C$

$$\varepsilon\eta^+ \cdot \text{grad}\,\Phi(y^*(t_c+0)) + o(\varepsilon) \geqslant 0$$

Upon division by ε, and by letting ε tend to zero we obtain, for $\varepsilon > 0$

$$\eta^+ \cdot \text{grad}\,\Phi(y^*(t_c+0)) \geqslant 0$$

and for $\varepsilon < 0$

$$\eta^+ \cdot \text{grad}\,\Phi(y^*(t_c+0)) \leqslant 0$$

Accordingly

$$\eta^+ \cdot \text{grad}\,\Phi(y^*(t_c+0)) = \sum_{\alpha=0}^{n} \lambda_\alpha(t_c+0)\eta_\alpha^+ = 0$$

Hence, from (1.17)

$$\sum_{\alpha=0}^{n} \lambda_\alpha^- \eta_\alpha^- = 0 \tag{1.18}$$

and since (1.18) holds for all $y^*(t_i) + \eta^i \in T^i$, we conclude that λ^- is perpendicular to T^c. As $\lambda_o^- = \lambda_o(t_c) = 1$, it follows that

$$\lambda^- = \lambda(t_c) \tag{1.19}$$

By letting t tend to t_c in (1.9) we deduce by continuity, for $t < t_c$

$$\min_{u\in K_u} H(\lambda(t_c), x^*(t_c), u) = -\lambda_n(t_c) \tag{1.20}$$

and by a similar argument for $t > t_c$, we have

$$\min_{u\in K_u} H(\lambda(t_c+0), x^*(t_c+0), u) = -\lambda_n(t_c+0) \tag{1.21}$$

Then, from (1.16) and (1.19)

$$\lambda_n(t_c) - \lambda_n(t_c+0) = \sum_{\beta=0}^{n} \left.\frac{\partial g_\beta(x,\mu^c)}{\partial x_n}\right|_{x=x^*(t_c)} \lambda_\beta(t_c+0)$$

and since $g_n(x,\mu) \equiv 0$

$$\lambda_n(t_c) - \lambda_n(t_c+0) = \sum_{\beta=0}^{n-1} \left.\frac{\partial g_\beta(x,\mu^c)}{\partial x_n}\right|_{x=x^*(t_c)} \lambda_\beta(t_c+0)$$

$$= \left.\frac{\partial H_c(x,\mu^c)}{\partial x_n}\right|_{x=x^*(t_c)} \qquad (1.22)$$

At last from (1.20), (1.21) and (1.22) part (iii) of Theorem 1.2 is proved.

Condition (iv) of Theorem 1.2 is deduced from $\lambda(t) = \operatorname{grad}\Phi(y^*(t))$ for $t \in [t_i, t_c)$ and $t \in (t_i, t_j]$, from which it follows that $\lambda_o(t) = 1$ for $t \in [t_i, t_c)$ and $t \in (t_c, t_j]$. Indeed, $\lambda_o(t_c) = 1$ by continuity as $t \to t_c$, $t < t_c$.

Now let us suppose that $t_c = t_i$, and that there does not exist a path $x^{**}(\cdot)$: $[t_i - \varepsilon, t_j] \to \overline{X}$, $\varepsilon > 0$, generated by s^*, which coincides with $x^*(\cdot)$ on $[t_i, t_j]$, and consider a solution $y^c(\cdot)$: $[t_c, t_c + \alpha] \to R^{n+1}$ of

$$\dot{y} = F(y,u), \qquad u \in K_u$$

such that $y^c(t_c) = y^*(t_c) = y^c$. For sufficiently small α this solution exists and is unique, as $y^*(t_c)$ is an interior point of $R \times X$ and $F(\cdot)$ is C^1 on $X \times U$.

It follows from Theorem 1.1 and (ii) of Assumption 1.5 that

$$y^c(t) \in (A/\Sigma(C)) \cup \Sigma(C) \quad \text{for all} \quad t \in [t_c, t_c + \alpha]$$

for sufficiently small α, and hence

$$\Phi(y^c(t)) \geqslant C \quad \text{for all} \quad t \in [t_c, t_c + \alpha] \qquad (1.23)$$

As a consequence of Assumption 1.4, Case (b), there exists a ball $B(x^*(t_c)) \subset X$ such that function $\Phi(\cdot)$ is of class C^1 on $R \times B(x^*(t_c))$. Consequently, for sufficiently small α, relation (1.23) result in

$$\operatorname{grad}\Phi(y^c) \cdot F(y^c,u) \geqslant 0 \quad \text{for all} \quad u \in K_u \qquad (1.24)$$

Letting $\lambda(t_c) = \operatorname{grad}\Phi(y^c)$, and taking account of the fact that $f_n(x,u) \equiv 1$, (1.24) rewrites

$$\lambda_n(t_c) + \min_{u\in K_u} \sum_{\alpha=0}^{n-1} \lambda_\alpha(t_c) f_\alpha(x^*(t_c),u) =$$

$$= \lambda_n(t_c) + \min_{u\in K_u} H(\lambda(t_c),x^*(t_c),u) \geq 0 \tag{1.25}$$

Hence, for $t_c = t_i$, (1.20) is to be replaced by

$$\min_{u\in K_u} H(\lambda(t_c),x^*(t_c),u) \geq -\lambda_n(t_c) \tag{1.26}$$

By similar arguments as in the proof of condition (iii) of Theorem 1.2, one can prove that (1.21) and (1.22) hold for $t_c = t_i$ provided that Assumption 1.4, Case (b), is satisfied. Then condition (v) of Theorem 1.2 is a direct consequence of (1.26), (1.21), and (1.22).

The proofs of the other conditions of Theorem 1.2 for $t_c = t_i$ are similar to the ones for $t_c \neq t_i$.

1.3. EXAMPLE

Let us consider the following example, which is a slightly modified version of the one given by J Case in (4).

The profit of the owner of a roadside inn, on some prescribed interval of time $t_i \leq t \leq T$ is a function of the number of strangers who pass by on the road each day, and on the number of times the inn is repainted during that period. The ability to attract new customers into the inn depends on its appearance, which is supposed to be indexed by a number x_1. During time intervals between paint jobs, x_1 decays according to the law

$$\dot{x}_1 = -kx_1 \qquad k = \text{constant} > 0 \tag{1.1)'$$

The total profit in the planning period $t_i \leq t \leq T$ is supposed to be

$$W(T) = A\int_{t_i}^{T} x_1(t)dt - \sum_{\alpha=1}^{N(T)} C_\alpha$$

where $N(T)$ is the number of times the inn is repainted, C_α, $\alpha = 1,\dots N(T)$, the cost of each paint job, and A a strictly positive constant. The owner of the inn wishes to maximize his total profit or, equivalently, to minimize $-W(T)$.

Let us suppose first that $0 \leq x_1 \leq 1$, and that each time the inn is repainted the index of appearance of the inn undergoes a jump from its previous value, say x_1, to its subsequent value $x_1^+ = x_1 + \mu(1-x_1)$ with $0 < \mu \leqslant 1$; that is, we shall suppose that the owner need not repaint the inn completely. In the intervals of time between paint jobs we shall let $\mu = 0$, and indeed $x_1^+ = x_1$. Moreover we shall suppose that the cost of a paint job indexed by μ is of the forme μC, where C is a strictly positive constant.

In that example we have

$$x = (x_1, x_2), \quad x_2 \equiv t$$
$$f_o(x,u) = -Ax_1$$
$$f_1(x,u) = -kx_1$$
$$g_o(x,\mu) = \mu C$$
$$g_1(x,\mu) = \mu(1-x_1)$$
$$K_\mu = [0,1]$$
$$\theta = \{ (x_1,t): \ 0 \leqslant x_1 \leqslant 1, \ t = T \}$$

Now, for mathematical reasons, let us further modify the problem of J.Case by replacing the condition $0 \leqslant x_1 \leqslant 1$ by $0 < x_1 < +\infty$, and the target θ by the new one $\{ (x_1,t): \ x_1 > 0, \ t = T \}$. Let

$$X = \{ (x_1,t): \ x_1 > 0, \ t < T \}$$

Let Π be the set of all functions defined on X with range in $[0,1]$, and Δ be the collection of all closed subsets of X. Here P is irrelevant as u does not appear in the arguments of $f_o(\cdot)$ and $f_1(\cdot)$.

In this paragraph, a path in R^2 generated by a strategy that satisfies the necessary conditions for optimality will be denoted by $x^*(\cdot)$, and the strategy itself which is a candidate to be an optimal one will be denoted by $(Y^*,\pi^*(\cdot))$, $Y^* \in \Delta$, $\pi^*(\cdot) \in \Pi$, as $p^*(\cdot)$ does not play any role in our example.

First let us consider the case of a path $x^*(\cdot): [t_i, T] \to \overline{X}$ which exhibits one and only one discontinuity point, at time t_c.

Case (a). Suppose that $t_c \neq t_i$.

Taking account of condition (iv) of Theorem 1.2, we have

$$H(\lambda,x,u) = -Ax_1 - \lambda_1 kx_1$$
$$H_c(x,\mu) = \mu C + \lambda_1(t_c+0)\mu(1-x_1) = \mu(C+\lambda_1(t_c+0)(1-x_1))$$

From Equations (1.1) and (1.2) we get

$$\dot{\lambda}_1 = A + k\,\lambda_1 \tag{1.2)'}$$

$$\lambda_1(t_c) = (1-\mu)\lambda_1(t_c+0) \tag{1.3)'}$$

Condition (i) of Theorem 1.2 is useless, as u does not appear in $H(\lambda_1, x, u)$. Condition (ii) of Theorem 1.2 leads us to considering three cases, namely

$$\text{Case 1}: \quad C + \lambda_1(t_c+0)(1-x_1^c) > 0 \tag{1.4)'}$$

$$\text{Case 2}: \quad C + \lambda_1(t_c+0)(1-x_1^c) < 0 \tag{1.5)'}$$

$$\text{Case 3}: \quad C + \lambda_1(t_c+0)(1-x_1^c) = 0 \tag{1.6)'}$$

and it follows from that condition (ii) that

$$\text{in Case 1} \qquad \mu = 0 \tag{1.7)'}$$

$$\text{in Case 2} \qquad \mu = 1 \tag{1.8)'}$$

$$\text{in Case 3} \qquad \mu \text{ is undetermined} \tag{1.9)'}$$

Condition (iii) of Theorem 1.2 is written

$$Ax_1^*(t_c+0) + k\lambda_1(t_c+0)x_1^*(t_c+0)$$

$$= Ax_1^*(t_c) + k\lambda_1(t_c)x_1^*(t_c) \tag{1.10)'}$$

Replacing $\lambda_1(t_c)$ in (1.10)' by its expression (1.3)' yields

$$(A + k\lambda_1(t_c+0))(x_1^*(t_c+0) - x_1^*(t_c))$$

$$+ k\mu\lambda_1(t_c+0)x_1^*(t_c) = 0 \tag{1.11)'}$$

By integrating equation (1.2)' and making use of the transversality condition $\lambda_1(T) = 0$ one obtains

$$\lambda_1(t_c+0) = -\frac{A}{k}(1 - e^{-k(T-t_c)}) \tag{1.12)'}$$

Moreover, from the definition of $g_1(\cdot)$, we have

$$x_1^*(t_c+0) - x_1^*(t_c) = \mu(1 - x_1^*(t_c)) \tag{1.13)'}$$

Putting (1.12)' and (1.13)' in (1.11)' gives

$$\mu(x_1^*(t_c) - e^{-k(T-t_c)}) = 0 \tag{1.14)'}$$

Hence

$$\mu \neq 0 \quad \Rightarrow \quad x_1^*(t_c) = e^{-k(T-t_c)} \tag{1.15)'}$$

which means that, for $t_c \neq t_i$, $x^*(t_c)$ lies on the curve $x = e^{-k(T-t)}$.

Now by replacing $\lambda_1(t_c+0)$ in (1.4)' - (1.6)' by its expression (1.12)' we define three regions, namely R_1, R_2, R_3, associated with Case 1, Case 2, Case 3, respectively; that is

$$\left.\begin{array}{lll} (x_1^c, t_c) \in R_1 & \Leftrightarrow & x_1^c > \varphi(t_c) \\ (x_1^c, t_c) \in R_2 & \Leftrightarrow & x_1^c < \varphi(t_c) \\ (x_1^c, t_c) \in R_3 & \Leftrightarrow & x_1^c = \varphi(t_c) \end{array}\right\} (x_1^c, t_c) \in X$$

where $\varphi(t_c) = 1 - \dfrac{kC}{A(1-e^{-k(T-t_c)})}$

Then from (1.7)' - (1.9)' it follows that

$$(x_1^c, t_c) \in R_1 \quad \Rightarrow \quad \mu = 0 \qquad (1.16)'$$

$$(x_1^c, t_c) \in R_2 \quad \Rightarrow \quad \mu = 1 \qquad (1.17)'$$

$$(x_1^c, t_c) \in R_3 \quad \Rightarrow \quad \mu \quad \text{undetermined} \qquad (1.18)'$$

Hence

$$\mu \neq 0 \quad \Rightarrow \quad x^*(t_c) \in R_2 \cup R_3 \qquad (1.19)'$$

and from (1.15)' and (1.19)', $x^*(t_c)$ lies on the portion of the curve $x = e^{-k(T-t)}$ all of whose points are in $R_2 \cup R_3$. The curves $x = \varphi(t)$ and $x = e^{-k(T-t)}$ have a single intersection point in X; let us denote it by $x = (\bar{x}_1, \bar{t})$. (1.16)' - (1.18)' together with (1.15)' implies that

for $t_c < \bar{t}$ $\quad\quad \mu = 1$, and

for $t_c = \bar{t}$ $\quad\quad \mu$ is undetermined.

Moreover there is no discontinuity for $t_c > \bar{t}$.

Case (b). Now let us treat the case $t_c = t_i$. (1.10)' is replaced by

$$Ax_1^*(t_c+0) + k\lambda_1(t_c+0)x_1^*(t_c+0) \geqslant Ax_1^*(t_c) + k\lambda_1(t_c)x_1^*(t_c) \qquad (1.20)'$$

which result in, by similar arguments as in case (a)

$$(A + k\lambda_1(t_c+0))(x_1^*(t_c+0) - x_1^*(t_c)) + k\mu\lambda_1(t_c+0)x_1^*(t_c) \geqslant 0 \qquad (1.21)'$$

then

$$\mu\,(x_1^*(t_c) - e^{-k(T-t_c)}) \leqslant 0 \qquad (1.22)'$$

and since $\mu \neq 0$ at a discontinuity point, (1.22)' implies that

$$x_1^*(t_c) \leqslant e^{-k(T-t_c)} \qquad (1.23)'$$

(1.16)' - (1.18)' hold, so that we have again

$$x^*(t_c) \in R_2 \cup R_3 \qquad (1.24)'$$

So, in Case (b), the conclusion is that $x^*(t_c)$ lies in the subset Y_1 of X

$$Y_1 = \{(x_1,t): \quad 0 < x_1 \leq \min(\varphi(t),\ e^{-k(T-t)}),\ t < T\}$$

If $x^*(t_c)$ is an interior point of Y_1, or lies on the portion of the boundary of Y_1 defined by

$$x = e^{-k(T-t)}, \quad x < \varphi(t)$$

it follows from (1.17)' that $\mu = 1$. If $x^*(t_c)$ lies on the portion of the boundary of Y_1 defined by

$$x = \varphi(t), \quad 0 < x \leq e^{-k(T-t)}$$

it follows from (1.18)' that μ is undetermined.

From now on we shall place a restriction on optimal strategies, in our example; that is, in addition to properties (i)-(iii) of Definition 1.6 we shall require that $\pi^*(\cdot)$ be continuous on Y^*. This requirement makes necessary to let $\mu = 1$ whenever the above discussion leaves μ undetermined.

Then, one can easily verify that, in the strip $X_1 \subset X$,

$$X_1 = \{(x_1,t): \quad x_1 > 0, \quad T - 2(T-\bar{t}) < t < T\}$$

the paths along which the necessary conditions of Theorem 1.2 are satisfied have at most one discontinuity point. From the discussion above it follows that a strategy $s^* = (Y^*,\pi^*(\cdot))$ that satisfies the necessary conditions of optimality on the strip X_1 is such that

$$Y^* = \{x = (x_1,t): \quad x \in Y_1 \cap X_1\}$$

and

$$\pi^*(x) = 0 \quad \text{for} \quad x \in X_1 - Y^*$$

$$\pi^*(x) = 1 \quad \text{for} \quad x \in Y^*$$

Now let us come to the case of a path $x^*(\cdot): [t_i,T] \to \bar{X}$ which exhibits more than one discontinuity point. Consider for instance the case of two discontinuity points at times $\bar{t}_c$ and $\bar{\bar{t}}_c$, $\bar{\bar{t}}_c < \bar{t}_c$. Let us suppose that $x^*(\bar{t}_c)$ is known, and that we wish to find $x^*(\bar{\bar{t}}_c)$. As $\bar{t}_c \neq t_i$ (1.10)' applies with t_c replace by $\bar{t}_c$. It follows from the preceeding discussion that $\bar{t}_c \leq \bar{t}$, and that $\pi^*(x^*(\bar{t}_c)) = 1$. Then, from (1.3)' we have

$$\lambda_1(\bar{t}_c) = \lambda_1(T) = 0 \qquad (1.25)'$$

and from (1.13)', with t_c replaced by $\bar{t}_c$

$$x^*(\bar{t}_c + 0) = 1 \qquad (1.26)'$$

so that (1.10)' rewrites

$$A + k\lambda_1(\bar{t}_c + 0) = Ax_1^*(\bar{t}_c) \qquad (1.27)'$$

By replacing $\lambda_1(\bar{t}_c+0)$ in (1.27)' by its expression (1.12)', with t_c replaced by $\bar{t}_c$ one obtains

$$Ae^{-k(T-\bar{t}_c)} = Ax_1^*(t_c) \qquad (1.28)'$$

Integrating (1.1)' gives

$$x_1^*(t) = x_1^*(T)e^{k(T-t)} \quad \text{for} \quad \bar{t}_c < t \leqslant T$$

so that, passing to the limit

$$x_1^*(\bar{t}_c+0) = x_1^*(T)e^{k(T-\bar{t}_c)} \qquad (1.29)'$$

Then, from (1.26)', (1.28)' and (1.29)' we deduce

$$x_1^*(\bar{t}_c) = x_1^*(T) \qquad (1.30)'$$

Because of (1.25)' and (1.30)', $x^*(\bar{\bar{t}}_c)$ is obtained by using the same procedure as for the one-discontinuity problem with the new target θ_c

$$\theta_c = \{(x_1,t): \quad x_1 > 0 \quad t = \bar{t}_c\}$$

and the new terminal point $x^*(\bar{t}_c)$.

If $\bar{\bar{t}}_c \neq t_i$ one can prove by the same argument as above that

$$x_1^*(\bar{\bar{t}}_c) = x_1^*(\bar{t}_c) = x_1^*(T)$$

Another property of $x^*(\bar{t}_c)$ is that it lies in the strip X_2

$$X_2 = \{(x_1,t): \quad x_1 > 0 \quad T-2(T-\bar{t}) < t \leqslant \bar{t}\}$$

This is because there does not exist a path $x^*(\cdot)$ with a single discontinuity point $x^*(\bar{t}_c)$ at a time $\bar{t}_c$, $\bar{t}_c \leqslant T-2(T-\bar{t})$, along which the necessary conditions of Theorem 1.2. are satisfied, as one can easily verify from the discussion of the one-discontinuity problem. Moreover $\bar{t}_c \leqslant \bar{t}$ as we have already noted.

A direct consequence of the procedure for obtaining $x^*(\bar{\bar{t}}_c)$ when $x^*(\bar{t}_c)$ is known is that

$$\bar{t}_c - \bar{\bar{t}}_c = T - \bar{t}_c \qquad (1.31)'$$

and that (1.14)' and (1.22)' are now replaced by

$$\mu(x_1^*(\bar{\bar{t}}_c) - e^{-k(\bar{t}_c-\bar{\bar{t}}_c)}) = 0 \quad \text{if} \quad \bar{\bar{t}}_c \neq t_i \qquad (1.32)'$$

$$\mu(x_1^*(\bar{\bar{t}}_c) - e^{-k(\bar{t}_c-\bar{\bar{t}}_c)}) \leqslant 0 \quad \text{if} \quad \bar{\bar{t}}_c = t_i \qquad (1.33)'$$

respectively.

From (1.31)', (1.32)' and (1.33)' it follows that if $\bar{\bar{t}}_c \neq t_i$, $x^*(\bar{\bar{t}}_c)$ is located on the curve

$$x_1 = e^{-\frac{k}{2}(T-t)}$$

and if $\bar{\bar{t}}_c = t_i$, $x^*(\bar{\bar{t}}_c)$ belongs to the region

$$0 < x_1 \leq e^{-\frac{k}{2}(T-t)}$$

We shall leave it to the reader to verify that $x^*(\bar{\bar{t}}_c)$ lies in the strip X_3

$$X_3 = \{(x_1,t): \quad x_1 > 0 \quad T-3(T-t) < t \leq T-2(T-\bar{t})\}$$

The arguments above may be repeated for paths exhibiting an arbitrary number of discontinuity points, and the results obtained for the two-discontinuity problem are easily generalized. One deduces from them a strategy $s^* = (Y^*, \pi^*(\cdot))$ that satisfies the necessary conditions of optimality on X. Y^* is the closed shaded area of Fig. 1.1, and

$$\pi^*(x) = 0 \quad \text{for} \quad x \in X - Y^*$$

$$\pi^*(x) = 1 \quad \text{for} \quad x \in Y^*$$

Let us compute $V^*(x)$ for $x \in X_1$. There are three kinds of paths satisfying the necessary conditions of Theorem 1.2, emanating from points of X_1, that is for initial time t_i in the interval $(T-2(T-\bar{t}),T)$, namely

(i) paths which are continuous on $[t_i, T]$, and

(ii) paths which have a discontinuity at a time $t_c \neq t_i$, and

(iii) paths which have a discontinuity at the initial time.

We shall denote by R_A, R_B, R_C the corresponding sets of initial states; that is, R_A is the set of all initial states in X_1 for which the policy, according to Theorem 1.2, is never to repaint; R_B is the set of initial states for which the policy is to repaint once before the end, but not at the initial time; and R_C is the set of initial states for which repainting at the starting point is necessary. By computing the cost in these three cases one finds easily

$$\text{for} \quad x \in R_A, \quad V^*(x) = -\frac{A}{k}\,x_1(1 - e^{-k(T-t)})$$

$$\text{for} \quad x \in R_B, \quad V^*(x) = -\frac{A}{k}\,(1 + x_1 - 2\sqrt{x_1 e^{-k(T-t)}}) + C$$

$$\text{for} \quad x \in R_C \quad V^*(x) = -\frac{A}{k}\,(1 - e^{-k(T-t)}) + C$$

$V^*(\cdot)$ is continuous on X_1. A limiting surface is shown on Fig. 1.2 .

We shall leave it to the reader to verify that Assumptions 1.1 and 1.3 - 1.6 are satisfied, except for Assumption 1.4 which is not satisfied for points belonging to the intersection of the closures of R_A and R_C, for $\bar{t} < t < T$. However, optimal paths emanating from such points can be obtained by direct computation of the cost of transfer to the target.

PART II : A CLASS OF DYNAMIC GAMES WITH PIECE - WISE CONTINUOUS TRAJECTORIES

2.1. PROBLEM STATEMENT

2.1.1. Strategies and Paths

Now let us suppose that the evolution of state $x \in X$ is influenced by two agents J_1 and J_2, whom we shall call players, through their choice of strategies s_1 and s_2 in prescribed strategy sets S_1 and S_2, respectively.

Let U, V, M, N be prescribed non-empty open subsets of R^{c_1}, R^{c_2}, R^{c_3}, R^{c_4}, respectively. Let K_u, K_v, K_μ, K_ν, be prescribed non-empty subsets of U, V, M, N, respectively. Let P, E, Π, Ξ, be prescribed sets of functions defined on X with range in K_u, K_v, K_μ, K_ν, respectively. Again let Δ be the collection of all closed† subsets of X .

Definition 2.1. The strategy sets of J_1 and J_2 are $S_1 = \Delta \times P \times \Pi$ and $S_2 = \Delta \times E \times \Xi$, respectively.

We shall denote by $(Y_1, p(\cdot), \pi(\cdot))$ and $(Y_2, e(\cdot), \varepsilon(\cdot))$ the elements of S_1 and S_2, respectively.

Now let $f(\cdot)$ and $g(\cdot)$ be prescribed functions of class C^1 on $X \times U \times V$ and $X \times M \times N$, respectively; namely

$$f(\cdot) : \begin{cases} X \times U \times V \to R^n \\ (x,u,v) \mapsto f(x,u,v) \end{cases}$$

$$g(\cdot) : \begin{cases} X \times M \times N \to R^n \\ (x,\mu,\nu) \mapsto g(x,\mu,\nu) \end{cases}$$

with $f = (f_1,\ldots f_n)$, $g = (g_1,\ldots g_n)$; and $f_n(x,u,v) \equiv 1$ and $g_n(x,\mu,\nu) \equiv 0$.

Definition 2.2. A strategy pair (s_1, s_2), $s_1 = (Y_1, p(\cdot), \pi(\cdot)) \in S_1$, $s_2 = (Y_2, e(\cdot), \varepsilon(\cdot)) \in S_2$ is admissible if and only if

$$x \in Y_1 \cup Y_2 \Rightarrow x + g(x, \pi(x), \varepsilon(x)) \in X - (Y_1 \cup Y_2)$$

Let S be the set of all admissible strategy pairs.

---

† in the topology induced by R^n on X .

Definition 2.3. A function $x(\cdot): [t_o, t_1] \to \overline{X}^{\dagger}$, defined on some interval of time $[t_o, t_1]$ is a path in R^n generated by $s = (s_1, s_2) \in S$, $s_1 = (Y_1, p(\cdot), \pi(\cdot))$, $s_2 = (Y_2, e(\cdot), \varepsilon(\cdot))$, from initial state $x^o \in X$ if and only if

(i) $x(t_o) = x^o$; and

(ii) $x(\cdot)$ is piece-wise continuous on $[t_o, t_1]$ - Let $T[t_o, t_1]$ denote the set of its continuity points - and

(iii) $x(t) = x(t-0)$ for $t \neq t_o$; and

(iv) $t \in T[t_o, t_1] \Rightarrow x(t) \in Y_1 \cup Y_2$ and

$x(t+0) = x(t) + g(x(t), \pi(x(t)), \varepsilon(x(t)))$; and

(v) for all t which are not discontinuity points, except possibly at $t = t_1$ $x(t) \in X - (Y_1 \cup Y_2)$; and

(vi) $x(\cdot)$ is differentiable, and

$$\frac{dx(t)}{dt} = f(x(t), p(x(t)), e(x(t)))$$

for all $t \in [t_o, t_1]$ except on a subset of $[t_o, t_1]$ at most denumerable.

It follows from (iii) that $t_1 \notin T[t_o, t_1]$.

2.1.2. Playability

We shall suppose that J_1 and J_2 desire to steer the state from a given initial state x^o to a state belonging to a prescribed target set, $\theta \subset \partial X$.

Definition 2.4. A path $x(\cdot): [t_o, t_1] \to \overline{X}$ generated by $s = (s_1, s_2) \in S$ from initial state $x^o \in X$ is a terminating path if and only if $x(t_1) \in \theta$.

Definition 2.5. A strategy pair $(s_1, s_2) \in S_1 \times S_2$ is playable at x^o, $x^o \in X$, if and only if it is admissible and it generates a terminating path from x^o.

Let $J(x^o)$ denote the set of all strategy pairs playable at x^o, and $I(x^o, s)$ the set of all terminating paths generated from x^o by a playable strategy pair $s \in J(x^o)$.

2.1.3. Cost of a Path

Let $f_o(\cdot)$ and $g_o(\cdot)$ be prescribed functions of class C^1 on $X \times U \times V$ and $X \times M \times N$, respectively; namely

---

† $\overline{X}$ denotes the closure of X in the topology of R^n.

$$f_o(\cdot) : \begin{cases} X \times U \times V & \to R \\ (x,u,v) & \mapsto f_o(x,u,v) \end{cases}$$

$$g_o(\cdot) : \begin{cases} X \times M \times N & \to R \\ (x,\mu,\nu) & \mapsto g_o(x,\mu,\nu) \end{cases}$$

Let $\tilde{\theta}_o(\cdot)$ be a prescribed function of class C^1 on some domain $D \subset R^n$ containing $\overline{X}$

$$\tilde{\theta}_o(\cdot) : \begin{cases} D & \to R \\ x & \mapsto \tilde{\theta}_o(x) \end{cases}$$

Let $\theta_o(\cdot)$ coincide with $\tilde{\theta}_o(\cdot)$ on θ, and $\theta_o(x) = 0$, for all $x \in X$.

The cost of a path $x(\cdot)$: $[t_o, t_1] \to \overline{X}$ generated by $s = (s_1, s_2) \in S$ from initial state $x^o \in X$ is

$$V(x^o, s_1, s_2, x(\cdot)) = \theta_o(x(t_1)) + \int_{t_o}^{t_1} f_o(x(t), p(x(t)), e(x(t)))dt + \sum_{t \in T[t_o,t_1]} g_o(x(t), \pi(x(t)), \varepsilon(x(t)))$$

2.1.4. Optimality of a Strategy Pairs

Let us suppose that the prescription for J_1 is to minimize the cost of transfer from an initial state to the target, while the prescription for J_2 is to maximize it.

Definition 2.6. A strategy pair $s^* = (s_1^*, s_2^*)$ is optimal on X if and only if

(i) s^* is playable at all $x^o \in X$; and

(ii) there exists a function $V^*(\cdot)$: $X \to R$ such that for all $x^o \in X$ and all $x(\cdot) \in I(x^o, s^*)$

$$V(x^o, s_1^*, s_2^*, x(\cdot)) = V^*(x^o); \text{ and}$$

(iii) for all $x^o \in X$, and all $(s_1^*, s_2) \in J(x^o)$, and all $(s_1, s_2^*) \in J(x^o)$, and all $\bar{x}(\cdot) \in I(x^o, s_1^*, s_2)$, and all $\bar{\bar{x}}(\cdot) \in I(x^o, s_1, s_2^*)$

$$V(x^o, s_1^*, s_2, \bar{x}(\cdot)) \leq V^*(x^o) \leq V(x^o, s_1, s_2^*, \bar{\bar{x}}(\cdot))$$

By letting $V^*(x^o) = \theta_o(x^o)$ for all $x^o \in \theta$, $V^*(\cdot)$ is defined on $X \cup \theta$.

2.2. NECESSARY CONDITIONS FOR AN OPTIMAL STRATEGY PAIR

2.2.1. Assumptions

Here we need

<u>Assumption 2.1.</u> $(\bar{p}(\cdot), \bar{e}(\cdot), \bar{\pi}(\cdot), \bar{\varepsilon}(\cdot)), (\bar{\bar{p}}(\cdot), \bar{\bar{e}}(\cdot), \bar{\bar{\pi}}(\cdot), \bar{\bar{\varepsilon}}(\cdot)) \in P \times E \times \Pi \times \Xi$, $\bar{x} \in X$ and $(p(\cdot), e(\cdot), \pi(\cdot), \varepsilon(\cdot))$ such that

$p(x) = \bar{p}(x),\ e(x) = \bar{e}(x),\ \pi(x) = \bar{\pi}(x),\ \varepsilon(x) = \bar{\varepsilon}(x)$ for $x_n < \bar{x}_n$

$p(x) = \bar{\bar{p}}(x),\ e(x) = \bar{\bar{e}}(x),\ \pi(x) = \bar{\bar{\pi}}(x),\ \varepsilon(x) = \bar{\bar{\varepsilon}}(x)$ for $x_n \geq \bar{x}_n$

implies that

$$(p(\cdot), e(\cdot), \pi(\cdot), \varepsilon(\cdot)) \in P \times E \times \Pi \times \Xi$$

Like in Part I, Assumption 2.1 assures that paths can be "joined".

<u>Assumption 2.2.</u> There exists a strategy pair (s_1^*, s_2^*) optimal on X, $s_1^* = (Y_1^*, p^*(\cdot), \pi^*(\cdot)) \in S_1$, $s_2^* = (Y_2^*, e^*(\cdot), \varepsilon^*(\cdot)) \in S_2$.

Let $Y^* = Y_1^* \cup Y_2^*$. Let $x^*(\cdot): [t_i, t_j] \to \bar{X}$ be a path generated by (s_1^*, s_2^*), such that $T\,[t_i, t_j] = t_c$.

<u>Assumption 2.3.</u> There exists an open ball $B(x^*(t_c+0) \subset X - Y^*$ with center $x^*(t_c+0)$, on which $p^*(\cdot)$, $e^*(\cdot)$ and $V^*(\cdot)$ are of class C^1.

<u>Assumption 2.4.</u> If $x^*(t_c) \in \partial Y^*$ there exists an open ball $B(x^*(t_c)) \subset X$ such that

(i) $p^*(\cdot)$ and $e^*(\cdot)$ agree on $(X - Y^*) \cap B(x^*(t_c))$ with functions - say $\bar{p}(\cdot)$ and $\bar{e}(\cdot)$ respectively - which are of class C^1 on $B(x^*(t_c))$; and

(ii) $\pi^*(\cdot)$ and $\varepsilon^*(\cdot)$ agree on $\partial Y^* \cap B(x^*(t_c))$ with functions - say $\bar{\pi}(\cdot)$ and $\bar{\varepsilon}^*(\cdot)$ respectively - which are of class C^1 on $B(x^*(t_c))$; and

(iii) $V^*(\cdot)$ is C^1 on $(X - Y^*) \cap B(x^*(t_c))$.

<u>Assumption 2.5.</u> The sets $\Omega_1(y^*(t_c))$ and $\Omega_2(y^*(t_c))$, namely

$$\Omega_1(y^*(t_c)) = \{y^*(t_c) + G(x^*(t_c), \pi^*(x^*(t_c)), \nu): \quad \nu \in K_\nu\}$$

$$\Omega_2(y^*(t_c)) = \{y^*(t_c) + G(x^*(t_c), \mu, \varepsilon^*(x^*(t_c))): \quad \mu \in K_\mu\}$$

are x_o^- and x_o^+ - directionally convex, respectively.

$G(\cdot) = (g_o(\cdot), g(\cdot))$, and $y^*(\cdot) = (x_o^*(\cdot), x^*(\cdot))$ is a path in R^{n+1} generated by (s_1^*, s_2^*), whose projection in R^n is $x^*(\cdot)$.

<u>Assumption 2.6.</u> If $x^*(t_c) \in \partial Y^*$ there exists an open ball $B(x^*(t_c)) \subset X$ in which the boundary of Y^* is smooth; that is

$$x \in \partial Y^* \cap B(x^*(t_c)) \Rightarrow h(x) = 0$$

where $h(\cdot): B(x^*(t_c)) \to R$ is of class C^1, and grad $h(x) \neq 0$ for all $x \in B(x^*(t_c))$.

2.2.2. Theorem 2.1.

Let $\lambda(\cdot): [t_i, t_j] \to R^{n+1}$ be a piece-wise continuous function, with $\lambda(t) = \lambda(t-0)$ for all $t \in (t_i, t_j]$, and let

$$H(\lambda,x,u,v) = \sum_{\alpha=0}^{n-1} \lambda_\alpha f_\alpha(x,u,v) \qquad \lambda = \lambda(t)$$

$$H_c(x,\mu,\nu) = \sum_{\alpha=0}^{n-1} \lambda_\alpha(t_c+0) g\ (x,\mu,\nu)$$

Theorem 2.1. If Assumptions 2.1 - 2.6 are satisfied and $t_c \neq t_i$, *there exist* $\tau > 0$ *and* $k \in R$ *and* $\lambda_\alpha(\cdot)$, $\alpha = 0,1,\ldots n-1$, *solutions of equations (2.1) on* $[t_c - \tau_1 t_c)$ *and* $(t_c, t_c+\tau]$

$$\dot{\lambda}_\alpha = - \left.\frac{\partial H(\lambda,x,u,v)}{\partial x_\alpha}\right|_{x=x^*(t),\ u=p^*(x),\ v=e^*(x)} \tag{2.1}$$

with

$$\lambda_\alpha(t_c) = \lambda_\alpha(t_c+0) + \left.\frac{\partial H_c(x,\mu,\nu)}{\partial x_\alpha}\right|_{x=x^*(t_c),\mu=\pi^*(x),\nu=\varepsilon^*(x)} + k \left.\frac{\partial h(x)}{\partial x_\alpha}\right|_{x=x^*(t_c)} \tag{2.2}$$

such that

(i) $$\min_{u\in K_u} H(\lambda(t),x^*(t),u,e^*(x^*(t))) = \max_{v\in K_v} H(\lambda(t),x^*(t),p^*(x^*(t)),v) =$$
$$= H(\lambda(t),x^*(t),p^*(x^*(t)),e^*(x^*(t)))$$

for all $t \in [t_c-\tau, t_c+\tau]$, $\quad t \neq t_c$; *and*

(ii) $$\min_{\mu\in K_\mu} H_c(x^*(t_c),\mu,\varepsilon^*(x^*(t_c))) = \max_{\nu\in K_\nu} H_c(x^*(t_c),\pi^*(x^*(t_c)),\nu) =$$
$$= H_c(x^*(t_c),\pi^*(x^*(t_c)),\varepsilon^*(x^*(t_c)));$$ *and*

(iii) $$\min_{u\in K_u} H(\lambda(t_c+0),x^*(t_c+0),u,e^*(x^*(t_c+0))) -$$
$$- \min_{u\in K_u} H(\lambda(t_c),x^*(t_c),u,e^*(x^*(t_c))) =$$
$$\max_{v\in K_v} H(\lambda(t_c+0),x^*(t_c+0),p^*(x^*(t_c+0)),v) -$$
$$- \max_{v\in K_v} H(\lambda(t_c),x^*(t_c),p^*(x^*(t_c)),v) =$$

$$= \left.\frac{\partial H_c(x,\mu,\nu)}{\partial x_n}\right|_{x=x^*(t_c),\mu=\pi^*(x),\nu=\varepsilon^*(t_c)} + k \left.\frac{\partial h(x)}{\partial x_n}\right|_{x=x^*(t_c)}$$

and

(iv) $\lambda_o(t) = 1$ *for all* $t \in [t_c - \tau,\ t_c + \tau]$

If $t_c = t_i$, *and if there exists an open ball* $B(x^*(t_c)) \subset X$ *on which* $\pi^*(\cdot)$ *and* $\varepsilon^*(\cdot)$ *are of class* C^1, *and if Assumptions 2.1 - 2.3 and 2.5 are satisfied, then (i) above is satisfied for all* $t \in (t_c,\ t_c + \tau]$, *and (ii) is satisfied, and (iv) is satisfied for all* $t \in [t_c,\ t_c + \tau]$, *and*

$$\lambda_\alpha(t_c) = \lambda_\alpha(t_c + 0) + \left.\frac{\partial H_c(x,\mu,\nu)}{\partial x_\alpha}\right|_{x=x^*(t_c),\mu=\pi^*(x),\nu=\varepsilon^*(x)}$$

$\alpha = 0,1,\ldots n-1$

and if $x^*(t_c) \notin Y_2^*$

(v) $$\min_{u \in K_u} H(\lambda(t_c + 0),\ x^*(t_c + 0),\ u,\ e^*(x^*(t_c + 0))) - \inf_{u \in K_u} H(\lambda(t_c),\ x^*(t_c),\ u,\ e^*(x^*(t_c)) \leq \left.\frac{\partial H_c(x,\mu,\nu)}{\partial x_n}\right|_{x=x^*(t_c),\mu=\pi^*(x),\nu=\varepsilon^*(x)}$$

and if $x^*(t_c) \notin Y_1^*$

(vi) $$\max_{v \in K_v} H(\lambda(t_c + 0),\ x^*(t_c + 0),\ p^*(x^*(t_c + 0)),\ v) - \sup_{v \in K_v} H(\lambda(t_c),\ x^*(t_c),\ p^*(x^*(t_c)),\ v) \geq \left.\frac{\partial H_c(x,\mu,\nu)}{\partial x_n}\right|_{x=x^*(t_c),\mu=\pi^*(x),\nu=\varepsilon^*(x)}$$

Theorem 2.1 needs be complemented by a *transversality condition*. By assuming that θ is a domain in the (n-1)-dimensional plane $x_n = T$, the transversality condition is

$$\lambda_\alpha(T) = \partial\theta_0(x)/\partial x_\alpha \Big|_{x = x^*(T)} \qquad \alpha = 1,\ldots n-1$$

2.3. EXAMPLE

Here we consider an example of worst case design in which nature imparts impulses to a system which is controlled by player J_1 otherwise. We shall suppose that the state of the system, $x = (x_1, x_2)$ with $x_2 \equiv t$, belongs to an open subset X of R^2, namely

$$X = \{x = (x_1, t): \quad 0 < x_1 < \alpha, \quad t < T\}$$

During time intervals between impulses, the motion of x is influenced by J_1 through the choice of a function $p(\cdot) \in P$, according to the control law

$$\dot{x}_1 = u\, x_1, \qquad u = p(x)$$
$$\dot{x}_2 = 1 \qquad (2.1)'$$

Let us suppose for instance that P is the set of all functions defined on X with range in $[0, a]$, which are C^1 with respect to x_1, and piece-wise continuous with respect to t. The target is

$$\theta = \{x = (x_1, t): \quad 0 < x_1 < \alpha, \quad t = T\}$$

The payoff, which player J_1 wishes to minimize, is supposed to consist of a terminal part and an impulsive part; that is, for any $p(\cdot) \in P$ and any initial state $x = (x_1^i, t_i)$ the payoff is

$$W(T) = -\, x_1(T) + \sum_{c=1}^{N(T)} \nu_c\, C \qquad C > 0$$

where N(T) is the number of impulses on $[t_i, T]$. Nature is replaced by a player J_2 which wishes to maximize W(T), and which controls the discontinuities by means of the equation

$$x_1(t+0) = x_1(t) + \nu\varphi(x_1(t), t), \qquad -1 < -b \leqslant \nu \leqslant 0$$

$\nu = \nu_c \neq 0$ at a discontinuity point t_c, $c = 1, \ldots N(T)$. Function $\varphi(\cdot): X \to R$ is prescribed. We shall suppose that

$$\varphi(x) = x_1\, e^{-\beta(x_1 - m)^2} \qquad \beta > 0, \qquad 0 < m < \alpha$$

As shown on Fig. 2.3 the derivative $\varphi'(x_1)$ of $\varphi(x_1)$ with respect to x_1 vanishes for $x_1 = x_{1M}$, and $\varphi'(x_1) > 0$ for $0 \leqslant x_1 < x_{1M}$, and $\varphi'(x_1) < 0$ for $x_{1M} < x_1$. We shall let $\varphi_M = \varphi(x_{1M})$, and we shall suppose from now on that

$$m < \alpha < \varphi_M$$

Also we shall suppose that

$$m < C < \alpha$$

In other words, in that example we have

$$f_o(x,u,v) = 0$$
$$f_1(x,u,v) = ux_1$$
$$g_o(x,\mu,\nu) = \nu C\,, \qquad m < C < \alpha, \qquad \theta_o(x(T)) = -x_1(T)$$
$$g_1(x,\mu,\nu) = \nu\varphi(x_1)$$
$$K_u = [0,a]\,, \quad K_\nu = [-b,\ 0] \text{ with } b < 1$$

First let us consider the case of a path $x^*(\cdot)$: $[t_i,\ T] \to \overline{X}$ which exhibits one discontinuity point only, at time t_c. Let $x^c = x^*(t_c)$.

Case (a). Suppose that $t_c \neq t_i$.

Taking account of Condition (iv) of Theorem 2.1. we have

$$H(\lambda,x,u,v) = \lambda_1 ux_1$$
$$H_c(x,\mu,\nu) = \nu C + \lambda_1(t_c+0)\nu\,\varphi(x_1)$$

From equations (2.1) and (2.2) we get

$$\dot{\lambda}_1 = -u\lambda_1 \qquad (2.2)'$$
$$\lambda_1(t_c) = (1+\nu\varphi'(x_1^c))\lambda_1(t_c+0) \qquad (2.3)'$$

From the transversality condition we deduce

$$\lambda_1(T) = -1 \qquad (2.4)'$$

Hence, there exists an interval $[T-\tau,\ T]$ on which $\lambda_1(t) < 0$, and from Condition (i) of Theorem 2.1, $u = a$ on that interval, so that

$$\lambda_1(t) = -e^{a(T-t)} \qquad (2.5)'$$

As $\lambda_1(t)$ is strictly negative on $[T-\tau,\ T]$, (2.5)' holds on $(t_c,\ T]$. Hence

$$\lambda_1(t_c+0) = -e^{a(T-t_c)} \quad . \qquad (2.6)'$$

Condition (ii) of Theorem 2.1 leads us to considering three cases, namely

Case 1 : $C + \lambda_1(t_c+0)\,\varphi(x_1^c) > 0$ (2.7)'

Case 2 : $C + \lambda_1(t_c+0)\,\varphi(x_1^c) < 0$ (2.8)'

Case 3 : $C + \lambda_1(t_c+0)\,\varphi(x_1^c) = 0$ (2.9)'

and condition (ii) yields

in Case 1 $\quad \nu = 0$ (2.10)'

in Case 2 $\quad \nu = -b$ (2.11)'

in Case 3 $\quad \nu$ is undetermined (2.12)'

By replacing $\lambda_1(t_c+0)$ in (2.9)' by its expression (2.6)' we obtain

$$C - e^{a(T-t_c)}\,\varphi(x_1^c) = 0$$

or equivalently

$$T - t_c = \frac{1}{a}\,\mathrm{Log}\,C - \frac{1}{a}\,\mathrm{Log}\,\varphi(x_1^c) \qquad (2.13)'$$

The curve defined by

$$T - t = \frac{1}{a} \operatorname{Log} C - \frac{1}{a} \operatorname{Log} \varphi(x_1) \qquad (2.14)'$$

separates X into three disjoint sets, namely R_1, R_2 and the curve R_3, such that

for $(x_1^c, t_c) \in R_1$ (2.7)' is satisfied, and

for $(x_1^c, t_c) \in R_2$ (2.8)' is satisfied, and

for $(x_1^c, t_c) \in R_3$ (2.9)' is satisfied.

However the fact that (x_1^c, t_c) is a discontinuity point is in contradiction with (2.7)' which implies that $\nu = 0$. Accordingly

$$(x_1^c, t_c) \in R_2 \cup R_3$$

Condition (iii) of Theorem 2.1 reduces to†

$$a\lambda_1(t_c + 0)x_1^*(t_c + 0) - a\lambda_1(t_c)x_1^*(t_c) = 0 \qquad (2.15)'$$

By replacing in (2.15)' $\lambda_1(t_c)$ by its expression (2.3)' and $x_1^*(t_c + 0)$ by

$$x_1^*(t_c + 0) = x_1^*(t_c) + \nu\varphi(x_1^*(t_c)) \qquad \nu \neq 0$$

we obtain

$$\varphi(x_1^c) - x_1^c \varphi'(x_1^c) = 0 \qquad (2.16)'$$

From which we deduce

$$x_1^c = m \qquad (2.17)'$$

We conclude that x^c lies on the intersection of the straight line defined by (2.17)' with the set $R_2 \cup R_3$. Let us denote by (m, t_m) its intersection with curve R_3 .

An example of a path $x^*(\cdot)$ that satisfies the necessary conditions of Theorem 2.1 in Case (a) is shown on Fig. 2.4. One can easily verify that the curve Γ defined by

$$x_1 = me^{a(t - t_m)}$$

is tangent to the curve R_3 at point (m, t_m) and that all of its points except (m, t_m) lie in R_1 .

Case (b). $t_c = t_i$

According to Condition (vi) of Theorem 2.1, in Case (b) condition (2.16)' is replaced by

$$\varphi(x_1^c) - x_1^c \varphi'(x_1^c) \geqslant 0 \qquad (2.18)'$$

from which we deduce

$$x_1^c \geqslant m \qquad (2.19)'$$

---

† One can see easily that $k = 0$, as the problem now is a control problem.

In that case x^c lies in the intersection of the strip

$$\{x = (x_1,t): \quad m \leqslant x_1 \leqslant \alpha, \quad t < T\} \text{ with } R_2 \cup R_3$$

Now consider the case of a path $x^*(\cdot): [t_i, T] \to \overline{X}$ which exhibits more than one discontinuity point. Consider the case of two discontinuity points at times $\bar{t}_c$ and $\bar{\bar{t}}_c$, $\bar{\bar{t}}_c < t_c$. As in the example of paragraph 1.3 we shall place a restriction on optimal strategies by requiring that $\pi^*(\cdot)$ be continuous on Y^*. This requirement makes necessary to let $\nu = -b$ whenever $C + \lambda_1(t_c+0)\,\varphi(x_1^c) = 0$, $t_c = \bar{t}_c$ or $t_c = \bar{\bar{t}}_c$.

Since $\nu = -b$ at point $x^*(\bar{t}_c)$, and $x^*(\bar{t}_c) = m$, and $\varphi'(m) = 1$, condition (2.3)' rewrites

$$\lambda_1(\bar{t}_c) = (1-b)\lambda_1(\bar{t}_c+0) \qquad (2.20)'$$

As $\lambda_1(\bar{t}_c+0) < 0$ and $b < 1$, $\lambda_1(\bar{t}_c)$ is strictly negative. By similar argument as above using Condition (i) of Theorem 2.1 one finds that

$$u = a \quad \text{on} \quad (\bar{\bar{t}}_c, t_c]$$

For $\bar{\bar{t}}_c \neq t_i$, it follows from Condition (iii) of Theorem 2.1, by the same argument as above, that

$$x_1^*(\bar{\bar{t}}_c) = x_1^*(\bar{t}_c) = m \qquad (2.21)'$$

Since $\bar{\bar{t}}_c$ is a discontinuity point, $\nu = -b$ at time $\bar{\bar{t}}_c$. Accordingly

$$x_1^*(\bar{\bar{t}}_c+0) = m(1-b)$$

By integrating equation (2.1)' on $[\bar{\bar{t}}_c, \bar{t}_c]$ one finds

$$m(1-b)e^{-a(\bar{\bar{t}}_c - \bar{t}_c)} = m \qquad (2.22)'$$

from which $\bar{\bar{t}}_c - \bar{t}_c$ is deduced.

Indeed we must verify that

$$C + \lambda_1(\bar{\bar{t}}_c+0)\,\varphi(x^*(\bar{\bar{t}}_c)) \leqslant 0 \qquad (2.23)'$$

By integrating equation (2.2)' with $u = a$ we obtain

$$\lambda_1(\bar{\bar{t}}_c+0) = \lambda_1(\bar{t}_c)e^{a(\bar{t}_c - \bar{\bar{t}}_c)} \qquad (2.24)'$$

Then, from (2.20)' and (2.24)' we have

$$\lambda_1(\bar{\bar{t}}_c+0) = (1-b)\lambda_1(\bar{t}_c+0)e^{a(\bar{t}_c - \bar{\bar{t}}_c)} \qquad (2.25)'$$

At last from (2.22)' and (2.25)' we get

$$\lambda_1(\bar{\bar{t}}_c+0) = \lambda_1(\bar{t}_c+0)$$

and since $x_1^*(\bar{\bar{t}}_c) = x_1^*(\bar{t}_c)$, $C + \lambda_1(\bar{t}_c+0)\,\varphi(x^*(\bar{t}_c)) \leqslant 0$ implies (2.23)'. Moreover

$$\lambda_1(\bar{\bar{t}}_c) = (1-b)\lambda_1(\bar{\bar{t}}_c+0) = \lambda_1(\bar{t}_c)$$

For $\bar{\bar{t}}_c = t_i$ it follows from Condition (vi) of Theorem 2.1 that

$$x_1^*(\bar{\bar{t}}_c) \geqslant m$$

The arguments above may be repeated for paths exhibiting an arbitrary number of discontinuity points, and the results obtained for the two-discontinuity problem are easily generalised. One deduces from them a strategy pair $s^* = (s_1^*, s_2^*)$, $s_1^* = (Y_1^*, p^*(\cdot), \pi^*(\cdot))$, $s_2^* = (Y_2^*, e^*(\cdot), \varepsilon^*(\cdot))$, that satisfies the necessary conditions of optimality on X.

$Y_1^* = \phi$ and Y_2^* is the closed shaded area of Fig. 2.4, and

$$p^*(x) = a \qquad \text{for } x \in X$$

$$\varepsilon^*(x) = 0 \qquad \text{for } x \in X - Y^*$$

$$\varepsilon^*(x) = -b \qquad \text{for } x \in Y^*$$

$\pi^*(\cdot)$ and $e^*(\cdot)$ are irrelevant as μ and ν do not enter the equations.

REFERENCES

(1) A. Blaquière, F. Gérard, and G. Leitmann, Quantitative and Qualitative Games, Academic Press Inc., 1969

(2) A. Blaquière, Quantitative Games: Problem Statement and Examples, New Geometric Aspects *in* The Theory and Application of Differential Games (J.D. Grote Ed.) D. Reidel Publishing Company, 1975.

(3) A. Blaquière, Vector-Valued Optimization in Multi-Player Quantitative Games *in* Multicriteria Decision Making (G. Leitmann, A. Marzollo Eds.) Springer-Verlag, 1975.

(4) J. Case, Impulsively Controlled D.G. *in* The Theory and Application of Differential Games (J.D. Grote Ed.) D. Reidel Publishing Company, 1975.

(5) V.K. Gorbunov, Differential Impulse Games, Izv. Akad. Nauk SSSR, Techn. Kibernet. 1973 n.4, 80-84; *translation* Engineering Cybernetics 11 (1973) n.4, pp.614-621.

(6) T.L. Vincent and J.D. Mason, Disconnected Optimal Trajectories, J.O.T.A. Vol.3, N°4, pp.263-281, 1969

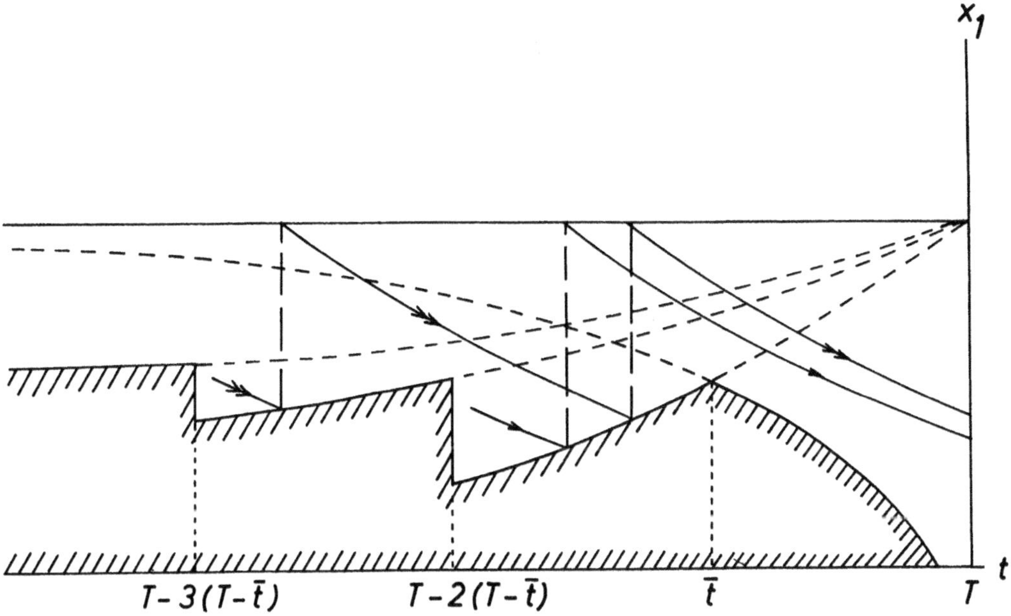

Fig. 1.1.

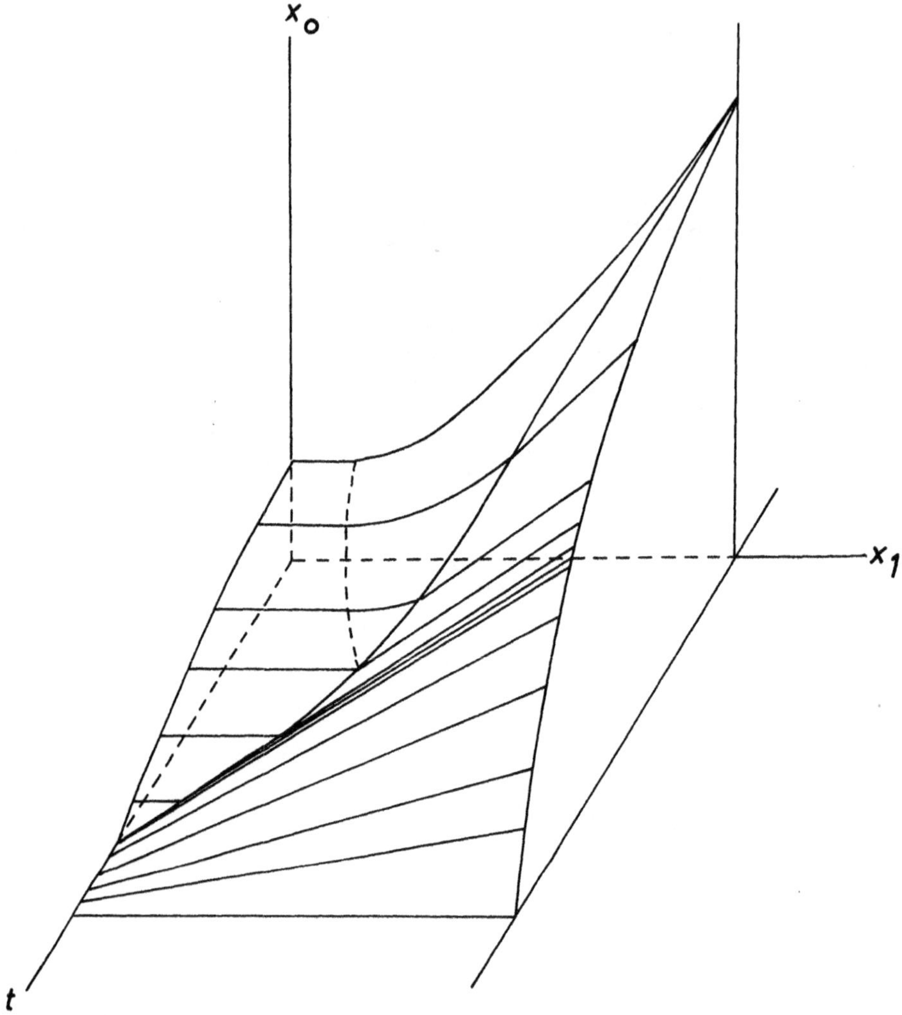

Fig. 1. 2.

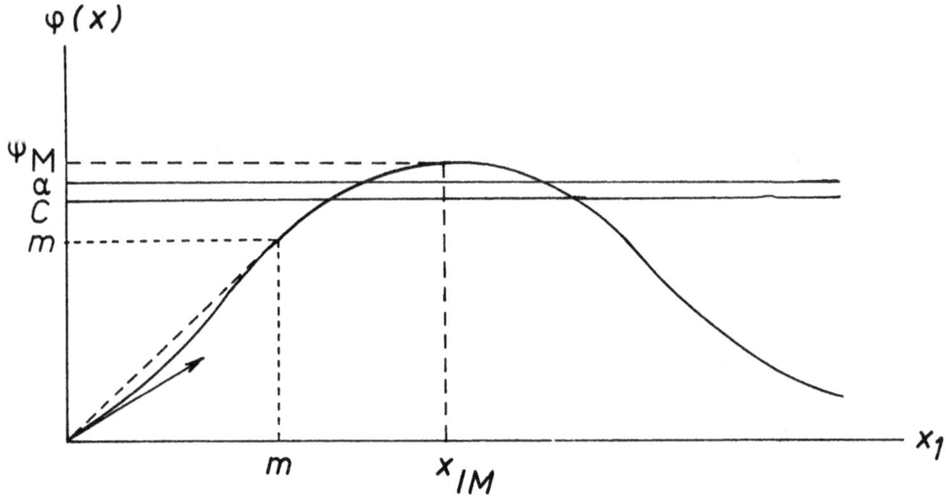

Fig. 2. 1.

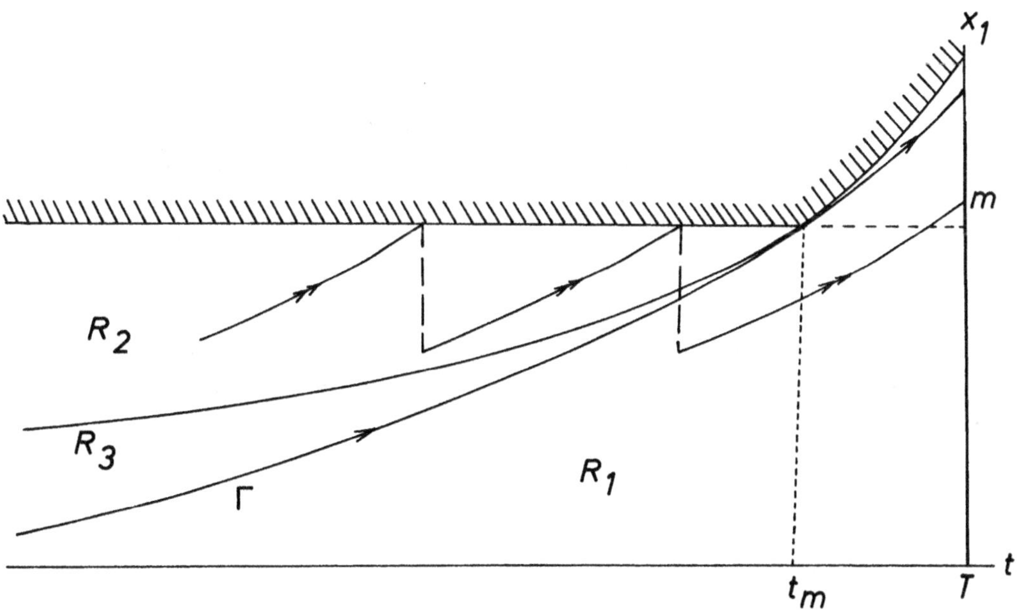

Fig. 2. 2.

LECTURE NOTES

John V. Breakwell, Professor
Department of Aeronautics & Astronautics
Stanford University
Stanford, California 94305/USA

A GENERAL PROBLEM

Given $\begin{cases} n \times 1 \\ \dot{x} = f(x,u,v) \\ \text{initial state } x_0 \end{cases}$, $\begin{cases} u \in U \\ v \in V \end{cases}$,

where the control sets U, V are independent of the state x, find the control histories u,v to be chosen by players 1 and 2, so as to minimize and maximize respectively the value of a certain function $M(x_f)$ of the "terminal state" x_f at the moment t_f when the state reaches a certain "terminal manifold" $C(x) = 0$. (Let us assume that all possible paths reach this terminal manifold in finite time.)

OUTLINE OF THE SOLUTION

Let $J(x)$ denote the value of $M(x_f)$ resulting from "optimal play" by both players when the initial state is x. Then (this is Isaacs' "main equation")

$$0 = \operatorname*{Min}_{u \in U} \operatorname*{Max}_{v \in V} J_x f(x,u,v) \left[= \operatorname*{Min}_{u \in U} \operatorname*{Max}_{v \in V} \frac{dJ}{dt} \right] \qquad (1)^{\ddagger}$$

thereby defining optimal controls (not necessarily unique) in the "closed-loop" form $u = u^*(x)$, $v = v^*(x)$. The optimality is an immediate consequence of (1): Player 1 by using $u^*(x)$, can prevent the path from moving off of the surface $J(x) = J(x_0)$ towards higher J-values; hence he can prevent $M(x_f)$ from exceeding $J(x_0)$. Similarly, player 2, by using $v^*(x)$, can prevent $M(x_f)$ from falling below $J(x_0)$. The determination of the optimal controls $u^*(x)$, $v^*(x)$ rests on the solution of the partial D.E. (1) for the function $J(x)$, subject to $J(x) = M(x)$ on $C(x) = 0$.

Suppose now that $J(x)$ is twice-differentiable inside some region $\mathcal{R}$. Partial differentiation of (1) w.r.t. x at any point in the interior of $\mathcal{R}$ shows that

$$0 = J_{xx} f(x,u^*(x),v^*(x)) + J_x(f_x + f_{u^*}u^*_x + f_{v^*}v^*_x) \; .$$

But $J_x f_{u^*} u^*_x = J_x f_{v^*} v^*_x = 0$, whether or not $u^*(x)$, $v^*(x)$ lie on the boundaries of sets U, V. Hence

$$\dot{J}_x \left(= J_{xx} f \right) = - J_x f_x \quad , \qquad (2)$$

as long as the optimal path remains in the interior of $\mathcal{R}$.

If $C(x) = 0$ is a smooth (n-1)-dimensional manifold, the terminal gradient

---

‡ We tacitly assume that the operations $\operatorname*{Min}_u$ and $\operatorname*{Max}_v$ commute.

$J_x(x_f)$ at any x_f is uniquely determined by (1) together with $J(x) = M(x)$ on $C(x) = 0$. The "adjoint equations" (2) may be integrated backwards from t_f along with the path equations $\dot{x} = f(x,u^*,v^*)$, with u^* and v^* determined by (1) in terms of x and J_x. In this way $J(x)$ and the optimal controls $u^*(x)$, $v^*(x)$ can be constructed in the interior of a region $\mathcal{R}$ bordered by $C(x) = 0$. (If $C(x) = 0$ is a manifold of dimension $\leq n-2$, the terminal gradient $J_x(x_f)$ and x_f will together still involve just $n-1$ free parameters, and the backwards construction may proceed.)

Remark. Typically, the state-space will be divided into regions $\mathcal{R}$ in which $J(x)$ is twice-differentiable, separated by "singular surfaces" across which the gradient J_x is discontinuous. The determination of such singular surfaces is crucial to the solution of most problems.

EXAMPLES:

Example 1 (cf. Isaacs [1], p. 76)

$$\begin{cases} \dot{x} = A(x,y)v - B(x,y)\sin u \\ \dot{y} = -1 - B(x,y)\cos u \end{cases}, \quad |v| \leq 1, \quad y_0 > 0$$

$C(y) \equiv y$ (the terminal surface is the x-axis); $M(x,y) = \text{some } M(x)$; suppose further $0 < B(x,y) < A(x,y) < 1$, so that termination, starting with $y_0 > 0$, is assured.

$$0 = \underset{u}{\text{Min}} \underset{\substack{v \\ |v|\leq 1}}{\text{Max}} \{J_x[A(x,y)v - B(x,y)\sin u] + J_y[-1 - B(x,y)\cos u]\}$$

$$\Rightarrow \begin{cases} v^* = \operatorname{sgn} J_x \\ (\sin u^*, \cos u^*) \parallel \text{ to } (J_x, J_y) \end{cases}$$

$$\Rightarrow A(x,y)|\sin u^*| = \cos u^* + B(x,y)$$

This result could have been obtained geometrically by observing that player 2 can dictate the sign of $\dot{x}$ to his advantage, and that player 1 therefore strives to minimize $\left|\frac{dx}{dy}\right|$. To do this he must choose $u = u^*$ so that (see Fig. 1) the direction $(\sin u^*, \cos u^*)$ is $\perp$ to the resultant $(\dot{x},\dot{y})$. This determines u^*:

$$A(x,y)|\sin u^*| - \cos u^* = B(x,y) \quad .$$

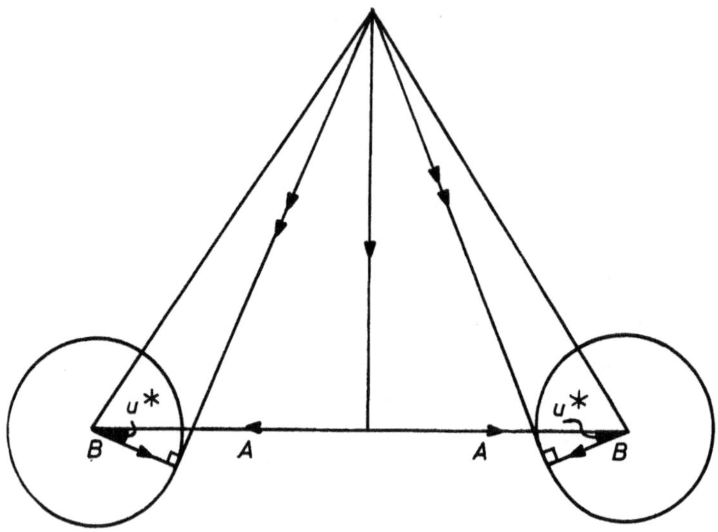

Figure 1.

On the x-axis, $J_x = M'(x)$, which determines the sign of $\sin u$. Figures 2 a,b,c, show the general appearance of the optimal paths for various types of function $M(x)$; in 2a, $M(x)$ has a single minimum, and the paths move right or left according as the position lies to the right or left of a certain "dispersal line," on which player 2 can choose either direction, with player 1 reacting as soon as he becomes aware of player 2's choice. Note that, in this two-dimensional state-space, the paths <u>are</u> the "surfaces" $J(x,y) = \text{constant}$, and that the gradient (J_x, J_y), which determines player 2's choice of direction $(\sin u^*, \cos u^*)$, is discontinuous across the dispersal line. In Fig. 2b, $M(x)$ has a single maximum, and there is a whole region (shaded) of positions from which player 2 can achieve $\text{Max } M(x)$. In Fig. 2c, $M(x)$ has one local maximum and two local minima; there are now three different dispersal lines as well as a finite region (shaded) from which player 2 secures the local maximum value of $M(x)$.

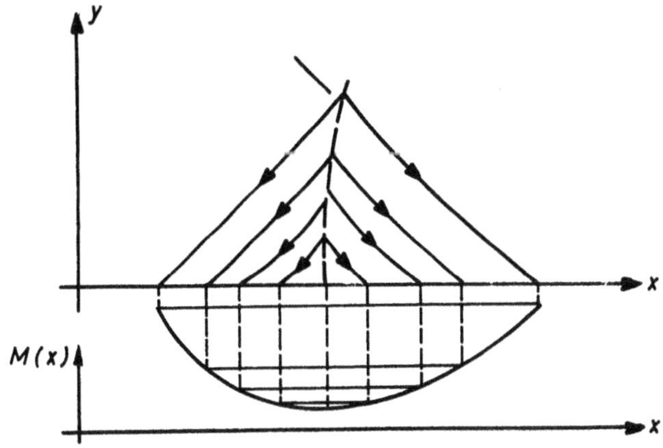

Figure 2a.

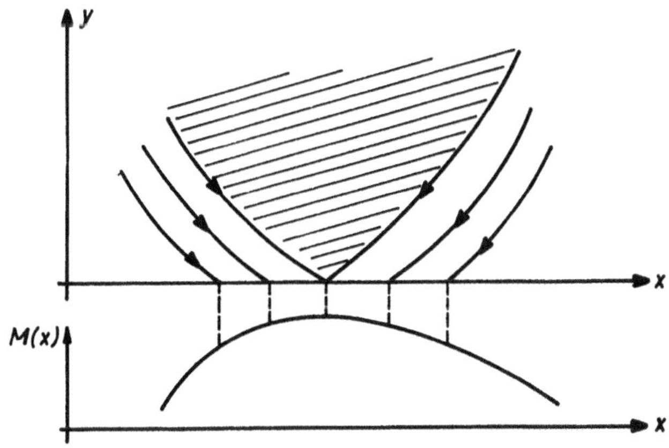

Figure 2b.

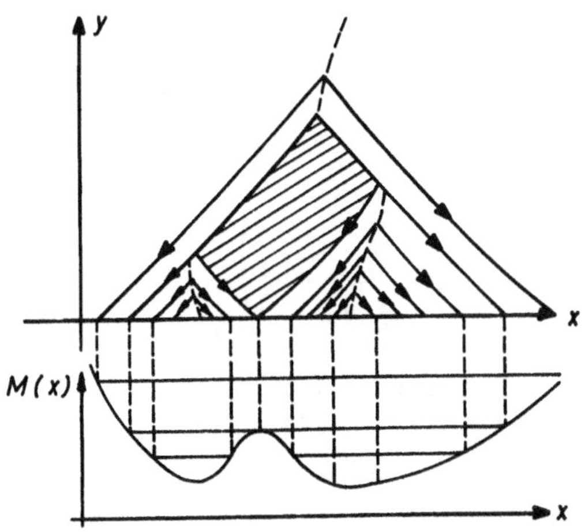

Figure 2c.

Example 2 "Lady in the Lake" ([1], reprinted edition, p. 386)

$$\begin{cases} P \text{ has speed 1, confined to the perimeter of a circle, radius } R \\ E \text{ has speed } \gamma < 1, \text{ swimming inside the circle.} \end{cases}$$

E wishes to land on the perimeter as far as possible from P. Figure 3 shows the appropriate coordinate system, θ denoting the angular position of E relative to P, $(-\pi \leq \theta \leq \pi)$, and the angle ψ determines E's direction.

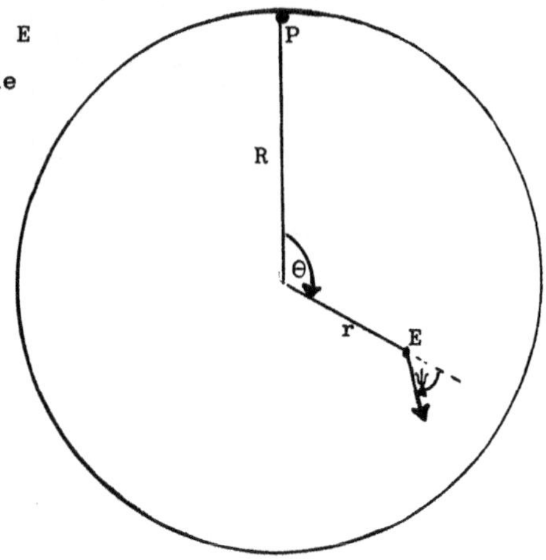

Figure 3.

$$\begin{cases} \dot{\theta} = \dfrac{\gamma \sin \psi}{r} - \dfrac{s}{R}, \quad s = \pm 1 \\ \dot{r} = \gamma \cos \psi \end{cases}$$

$$M = |\theta_f| .$$

$$0 = \underset{|s|=1}{\underset{s}{\text{Min}}} \ \underset{\psi}{\text{Max}} \left\{ J_r \gamma \cos \psi + J_\theta \left(\frac{\gamma \sin \psi}{r} - \frac{s}{R} \right) \right\}$$

$$\Rightarrow \begin{cases} s^* = \text{sgn } J_\theta = \text{constant} = \text{sgn } \theta_f \\ (\cos \psi^*, \sin \psi^*) \parallel \text{ to } (J_r, \frac{1}{r} J_\theta) \end{cases}$$

$$\Rightarrow \gamma - \frac{r}{R} \sin \psi^* \text{sgn } \theta_f = 0$$

Hence E runs in a straight line tangent to a circle of radius γR (see Fig. 4). This result could have been obtained geometrically by noting that E's motion in the

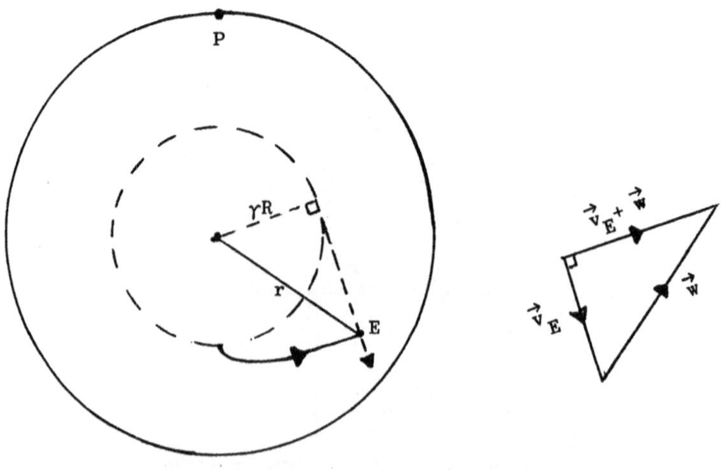

Figure 4.

relative coordinate system attached to P is the vector sum of his own velocity $\vec{v}_E$ and a velocity $\vec{w}$ obtained by rotation about the center, opposite to P's motion, with angular velocity $\frac{1}{R}$. As in Example 1, the optimal direction for $\vec{v}_E$ is $\perp$ to the resultant velocity $\vec{v}_E + \vec{w}$. Since $\vec{w}$ is $\perp$ to E's radius vector, and since $|\vec{v}_E|/|\vec{w}| = \frac{\gamma R}{r}$, E's velocity must be along a tangent to the γR-circle, and the resultant path, whose tangent is along $\vec{v}_E + \vec{w}$, is an evolute of this circle.

This construction for E's optimal direction fails if E lies inside the γR-circle. However, E can always start by moving to the relative position $\theta = \pi$ first inside and later on the γR-circle, inside of which she has angular advantage. Strictly speaking, E, not knowing P's instantaneous choice of direction, cannot maintain $\theta = \pi$, even inside the γR-circle. However, she can keep within an arbitrarily small angle ϵ of the position $\theta = \pi$ as she moves outward toward $r = \gamma R$. From the position $r = \gamma R$, $\theta = \pi$ she then breaks right or left along the tangent. Assuming that $\gamma > .217$, she does achieve a positive terminal angular separation: $\theta_f = \pi + \text{Cos}^{-1}\gamma - \frac{1}{\gamma}\sqrt{1-\gamma^2}$.

From some starting positions E can do better (see Fig. 5). The right and left evolutes intersect on $\theta = \pi$, a dispersal line. The shaded region consists of positions from which E must return to the inside of the γR-circle.

James Flynn has considered a related game "Lion and Man," with Lion, the pursuer, slower than Man, the evader, and both confined to a circular arena. In particular, if Man is constrained to remain on the perimeter of the arena [2], the problem is essentially the same as "Lady in the Lake;" E, in trying to reach the shore as far as possible from P, is in effect trying to come as close as possible to the position P' opposite to P. Now, however, the closest approach occurs before L reaches the perimeter (see Fig. 6), in fact, on the tangent from M to the γR-circle, this distance of closest approach being $D = R(\sqrt{1-\gamma^2} - \gamma\text{Cos}^{-1}\gamma)$. Again, L can do better than D from some starting positions, and the cross-hatched region indicates positions which are initially closer than D but from which the distance increases immediately.

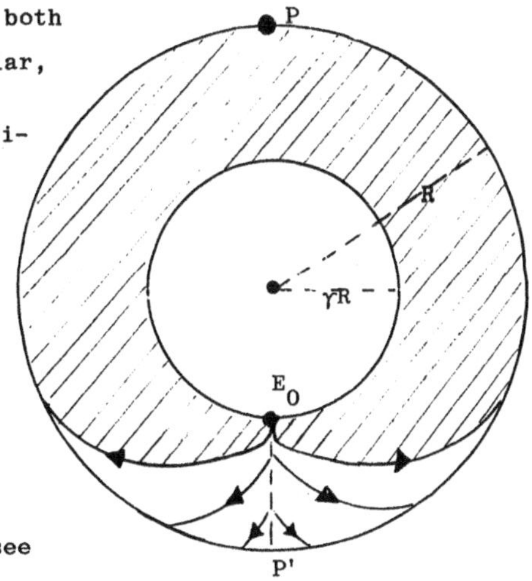

Figure 5.

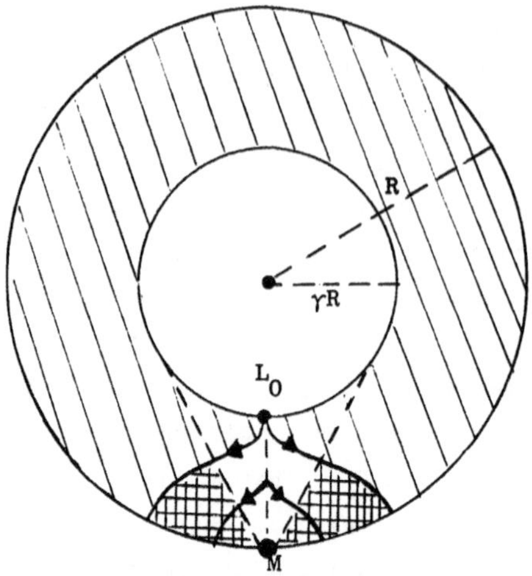

Figure 6.

<u>Example 3</u> <u>"Homicidal Chauffeur" and Related Games (in the plane)</u>

Suppose $\begin{cases} \text{P has speed 1 and maximum turn-rate 1} \\ \text{E has speed } \gamma < 1 \text{ and can change direction at will} \end{cases}$

P wishes to come as close as possible to E, who wishes to maximize the distance of closest approach. (This is really the "Game of Kind" in Isaacs' terminology, as opposed to the "Game of Degree" in which the payoff is the time taken to come within a specified capture distance.)

If $\vec{r}$ denotes the position of E relative to P, $\hat{u}_E$, $\hat{u}_P$ denote unit-vectors $\parallel$ to the motions of E,P, α denotes the direction of $\hat{u}_P$, measured from some fixed reference direction, ω_P denotes P's turn-rate, then:

$$\begin{cases} \dot{\vec{r}} = \gamma\hat{u}_E - \hat{u}_P(\alpha) \\ \dot{\alpha} = \omega_P \qquad , \quad |\omega_P| \leq 1 . \end{cases}$$

The main equation (1) becomes

$$0 = \min_{\substack{\omega_P \\ |\omega_P| \leq 1}} \max_{\hat{u}_E} \left\{ J_{\vec{r}} \cdot (\gamma\hat{u}_E - \hat{u}_P(\alpha)) + J_\alpha \omega_P \right\}$$

$$\Rightarrow \begin{cases} \hat{u}_E^* \text{ is } \parallel \text{ to } J_{\vec{r}} , \\ \omega_P^* = - \operatorname{sgn} J_\alpha \end{cases}$$

The adjoint equations (2) show that the gradient $J_{\vec{r}}$ is constant implying that E runs in a straight line (at least, as long as the path remains inside a region $\mathcal{R}$ in which J is twice-differentiable).

The motion can be described in a relative coordinate frame centered at P and oriented along P's direction of motion. If, in this frame, E's bearing from P is θ and E's heading is H (see Fig. 7), then

$$\begin{cases} \dot{r} = -\cos\theta + \gamma\cos(H-\theta) \\ \dot{\theta} = \dfrac{\sin\theta}{r} - \omega_P + \dfrac{\gamma}{r}\sin(H-\theta) \end{cases}$$

Figure 7.

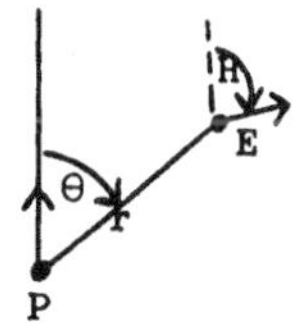

The main equation in these variables shows that $\omega_P^* = \text{sgn } J_\theta$ and $H^* = \theta + \text{Tan}^{-1}(J_\theta/rJ_r)$. ω_P^* is piecewise constant, equal to ± 1 (unless J_θ were to remain zero for some interval of time), while H^* in this coordinate system is not constant, but $(\sin H^*, \cos H^*)$ rotates at a rotate opposite to the rate ω_P of the frame.

In this game $M \equiv r$, so that at termination: $J_r = 1$, $J_\theta = 0$, $H_f = \theta_f$. The game ends at $\dot{r} = 0$; i.e., $C \equiv \dot{r}|_{H=\theta} = \gamma - \cos\theta$; hence $\theta_f = \pm \text{Cos}^{-1}\gamma$.

The paths may be constructed backwards for any final separation r_f; J_θ is found to have the same sign as θ_f immediately prior to termination. Hence P turns towards E who runs in a straight line which is directly away from P at closest approach. The paths of E in the relative frame may be easily constructed. Alternatively, we may note that the relative coordinate frame is rotating at unit angular rate, clockwise about the right turn-center C_+ or counter-clockwise about the left turn-center C_-. This angular rate is the same as in the game "Lady in the Lake" if R, there, is taken to be unity. Hence the paths in the relative frame are again evolutes of circles, now having radius γ and centered at C_+ or C_-, according as P is turning right or left ([1],[3]). This construction clearly fails if $r_f^2 + \gamma^2 > 1$. This case will be discussed later.

Figure 8 shows the paths for various r_f, with $r_f^2 + \gamma^2 < 1$, all terminating at $\theta_f = \pm \text{Cos}^{-1}\gamma$. If $r_f \leq \gamma \text{Sin}^{-1}\gamma + \sqrt{1-\gamma^2} - 1$, the paths originate on the

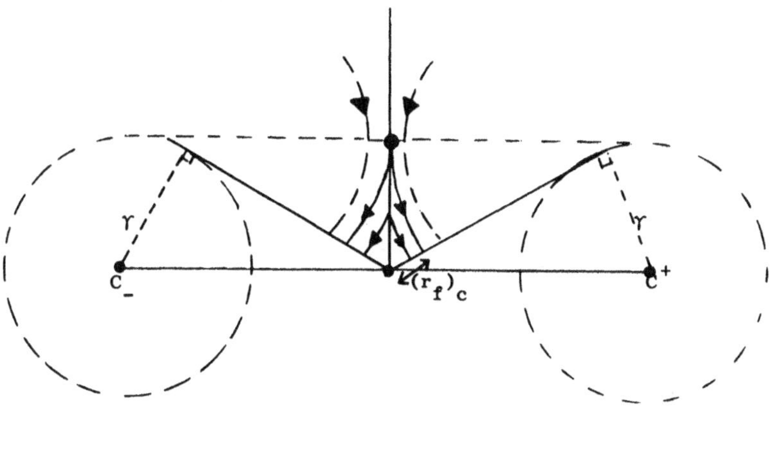

Figure 8.

dispersal line $\theta = 0$, on which E has a choice as to whether to break right or left. There is a finite "capture region corresponding to any such r_f: if E is outside this region he can prevent approach to within less than r_f, simply by running ⊥ to the contour for this r_f whenever the relative position reaches this contour. However, E cannot prevent approach to within any r_f exceeding the critical value $(r_f)_c = \gamma \text{Sin}^{-1}\gamma + \sqrt{1-\gamma^2} - 1$, since P can move E onto the line $\theta = 0$ and hold him there until $r = \gamma$ by first-of-all, if necessary, running away to sufficiently large r. $(r_f)_c$ is both the closest approach that P can achieve from <u>all</u> starting positions and the largest separation that E can guarantee from all starting positions except those within a finite capture region in front of P. Note that E's optimal strategy, assuming that $\theta = 0$ for $r > \gamma$, is to wait until $r = \gamma$ and then break right or left perpendicular to the relative position at this instant, at which $J_r = 0$.

Figure 9 shows the capture region for a rectangular car, assuming a typical relatively large turn-radius.

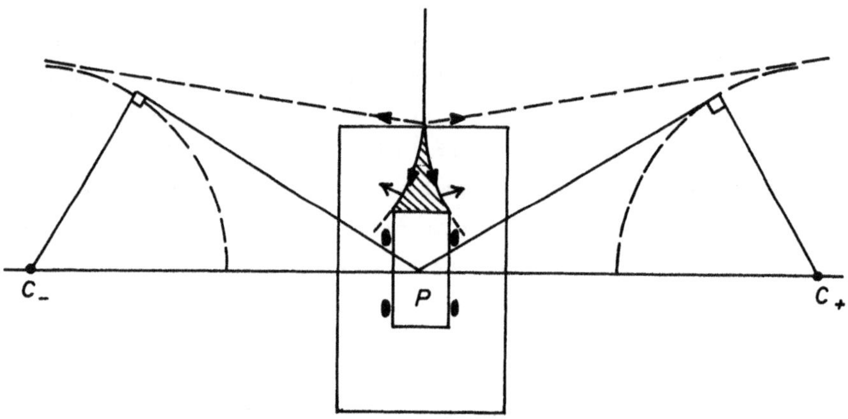

Figure 9.

Figure 10 shows the capture region for a similar rectangular car for the game "Suicidal Pedestrian," in which E is now trying to be captured and P to prevent capture. ω_P is now $-J_\theta$, but J_θ changes sign from + to - shortly before termination of the grazing path at a rear corner of the car; P turns initially away from E, but near termination he turns towards E to prevent the rear corner of the car from striking P.

Figures 11 a,b,c show the Surveillance Region in the game "Surveillance Evasion" [4], in which P strives to minimize the maximum separation r_f and E to maximize it. The Surveillance Region, for given r_f, indicates those positions from which the maximum separation does not exceed r_f, so that P can keep E within a surveillance range r_f. Figure 11a is applicable if $f_3(\gamma) \triangleq 1 + \sqrt{1-\gamma^2} + \gamma(\pi+\mathrm{Sin}^{-1}\gamma) < r_f$. Figure 11b is applicable if $f_2(\gamma) \triangleq \sqrt{1-\gamma^2} + \gamma(\pi+2\mathrm{Sin}^{-1}\gamma) < r_f < f_3(\gamma)$. Here J_θ changes sign at the point A, and the right barrier, for example, starts with a left turn by P, followed at A by a right turn.

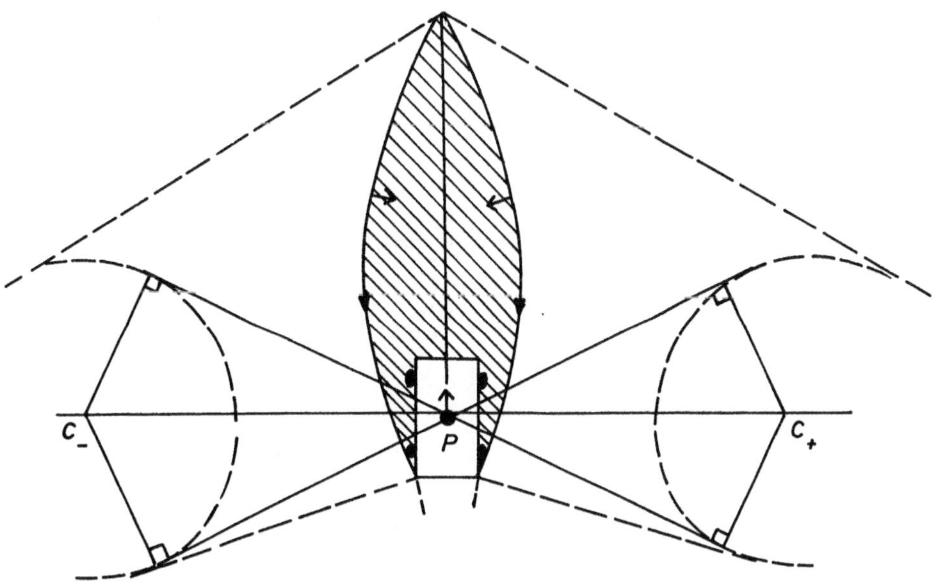

Figure 10.

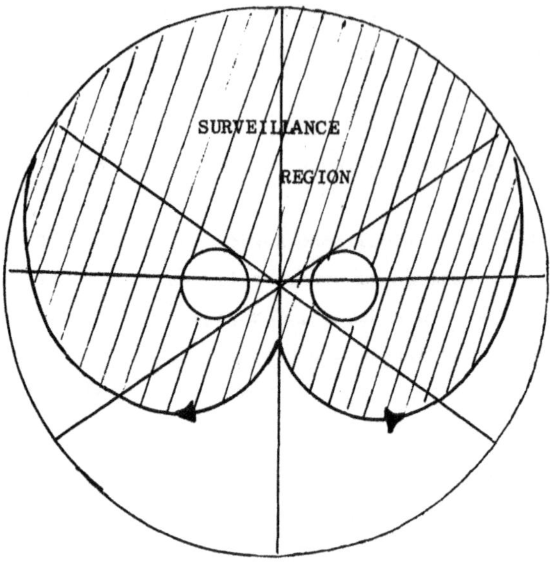

Figure 11a.

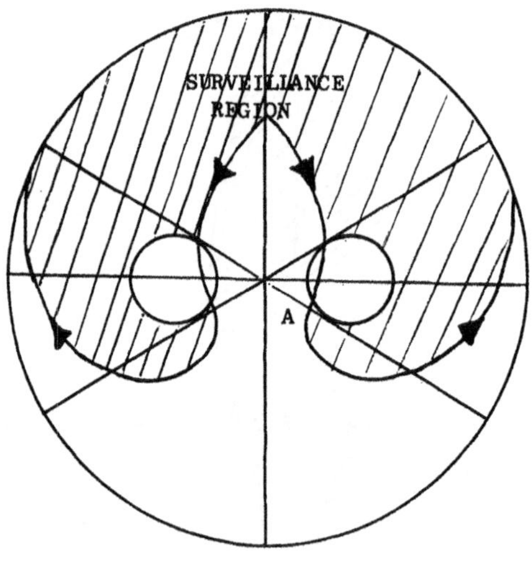

Figure 11b.

Lastly, if $f_1(\gamma) \stackrel{\Delta}{=} \sqrt{1-\gamma^2} + \gamma(1 + \frac{\pi}{2} - 2\mathrm{Sin}^{-1}\gamma) < r_f < f_2(\gamma)$, the Surveillance Region takes the form illustrated in Fig. 11c. Here P's left turn is followed at B by a right turn, but the barrier direction is now discontinuous at B. Since J_θ approaches zero as the path approaches B (in forward time), E cannot escape through this composite barrier [5].

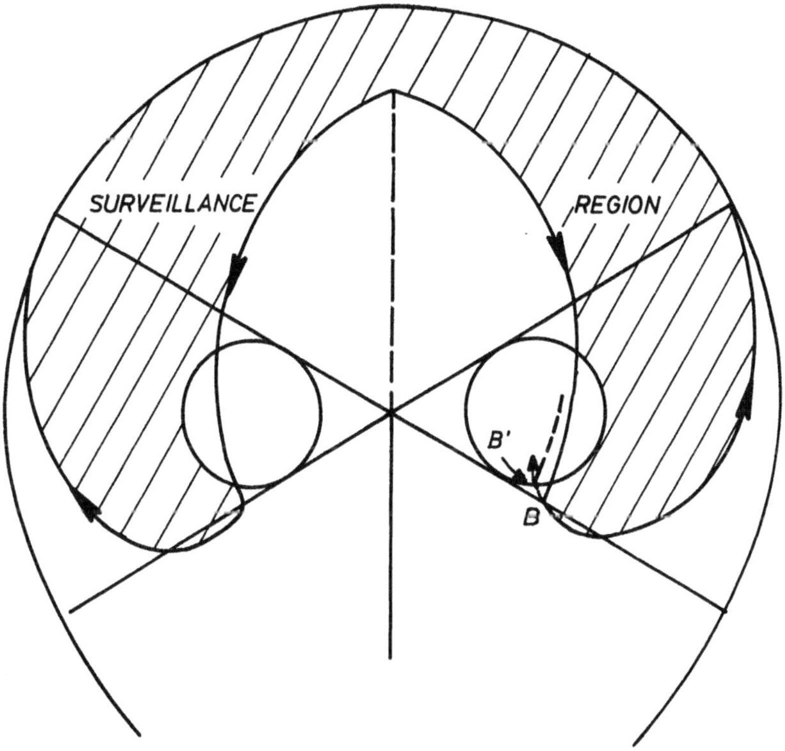

Figure 11c.

If, on the other hand, we attempt to define a larger Surveillance Region, bounded by a composite with a corner at B' in Fig. 11c, so that $J_\Theta(t^-_{B'}) \neq 0$, a switch to a right turn by P, together with a continuation of E's prior direction ⊥ to the arriving barrier, will lead to $\frac{d}{dt}(J^-) > 0$ and $\frac{d}{dt}(J^+) < 0$, implying a direction at B' (indicated by a dotted arrow) leading to escape from the hoped-for Surveillance Region.

Returning to the "Homicidal Chauffeur" game, we look finally at the case when $r_f^2 + \gamma^2 > 1$. It follows from the main equation that as $\dot{r} \to 0$, either $J_\Theta \to 0$ or $\dot{\Theta} \to 0$. As shown by Isaacs [1], it is not possible to construct a path with $[J_\Theta]_{t_f} = 0$ if $r_f^2 + \gamma^2 > 1$, since this leads to $\ddot{r}|_{t_f} < 0$. In this case, therefore, the terminal condition is one of relative equilibrium $\dot{r} = \dot{\Theta} = 0$. For positive Θ_f, for example, this requires $H_f > \Theta_f > \mathrm{Cos}^{-1}\gamma$. In the terminal configuration, P and E are describing concentric circles, at the same angular rate, of radii 1 and γ respectively. The critical capture radius $(r_f)_c$, moreover, is again reached by E running ⊥ to the initial position vector along $\Theta = 0$, since E optimizes the range for his initial break to one side or the other, implying that J_r is initially zero. Figure 12 shows the paths of both P and E in this case; E has moved unit distance in reaching equilibrium, so that P has turned through an angle $(1/\gamma)$. Hence $(r_f)_c = \sqrt{1+\gamma^2-2\gamma\sin(1/\gamma)}$.

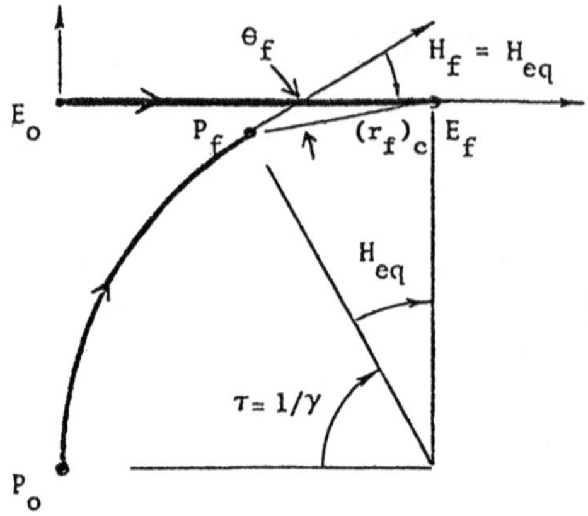

Figure 12.

Example 4 "Game of Two Cars" ([1], p.237)

Here the evader of E of example 3 also has a maximum turn-rate, say ω, with $\omega > 1$. The heading H now becomes a third state variable, and the complete equations are:

$$\begin{cases} \dot{r} = -\cos\theta + \gamma\cos(H-\theta) \\ \dot{\theta} = \dfrac{\sin\theta}{r} - \omega_P + \dfrac{\gamma}{r}\sin(H-\theta) \\ \dot{H} = \omega_E - \omega_P \end{cases} , \qquad \begin{cases} (\omega_P) \leqq 1 \\ (\omega_E) \leqq \omega \end{cases} .$$

The optimal turn-rates are $\omega_P^* = \text{sgn}(J_\theta + J_H)$, $\omega_E^* = \omega \text{ sgn } J_H$ as long as $J_\theta + J_H$ and J_H do not remain zero. The optimal paths of example 3, for which $\omega_E = 0$ and $\omega_P = \pm 1$, are again optimal, being now "singular arcs" along which J_θ remains zero. However, from most initial states (r,θ,H) a singular arc must be preceded by a "tributary arc" on which $|\omega_E| = \omega$.

In particular, the critical capture radius is now reduced to

$$(r_f)_c = \gamma \text{ Sin}^{-1}\gamma + \sqrt{1-\gamma^2} - 1 - \frac{\gamma}{\omega}\left(\frac{\pi}{2} - 1\right)$$

if (case I) this doesn't exceed $\sqrt{1-\gamma^2}$, as is evident from Fig. 13; E loses time in turning from a neutral initial heading $H = 0$ or $H = \pi$ towards his optimal direction which is again perpendicular to the initial relative position.

If $(r_f)_c$ exceeds $\sqrt{1-\gamma^2}$ (case II) the situation shown in Fig. 12 must be similarly modified, as shown in Fig. 14. The critical capture radius is now:

$$(r_f)_c = \sqrt{1 + \gamma^2 - 2\gamma \sin\left[\frac{1}{\gamma} + \left(\frac{\pi}{2} - 1\right)/\omega\right]} .$$

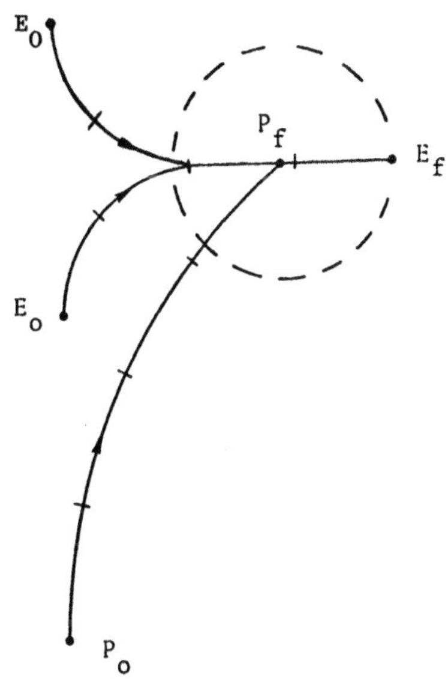

Figure 13 Critical Trajectories in Region I

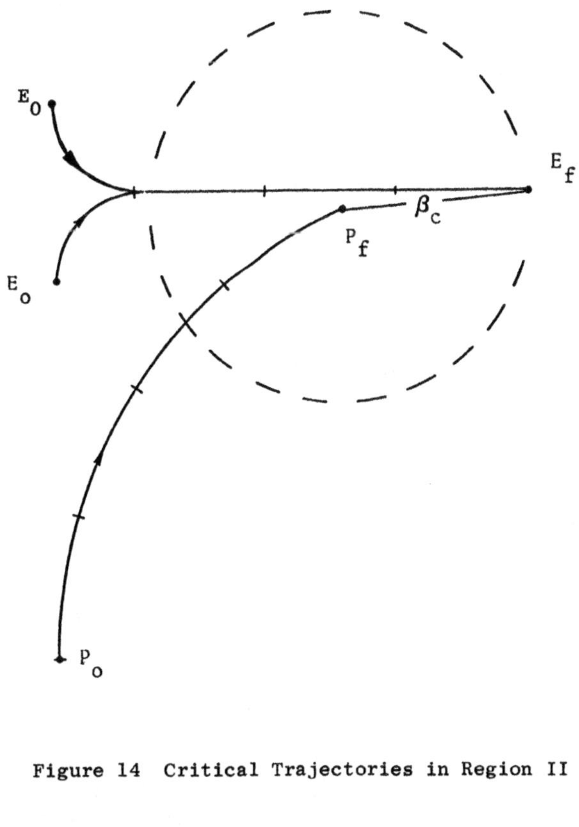

Figure 14 Critical Trajectories in Region II

Finally, if $\frac{1}{\omega} \leq \gamma < \sin\frac{\pi}{2\omega}$ (Case III), the minimum range occurs while E is still turning toward his optimal direction, i.e., there is no singular arc, and the following equations hold (see Fig. 15):

$$(r_f)_c + 1 - \cos\tau = \frac{\gamma}{\omega}(1 - \cos\omega\tau)$$

$$\sin\tau = \gamma \sin\omega\tau \quad ,$$

since $\dot{r}_f = 0$ and, as in Case I, $(J_\Theta)_{t_f} = 0$, $H_f = \Theta_f$, and the final relative position is perpendicular to the initial one.

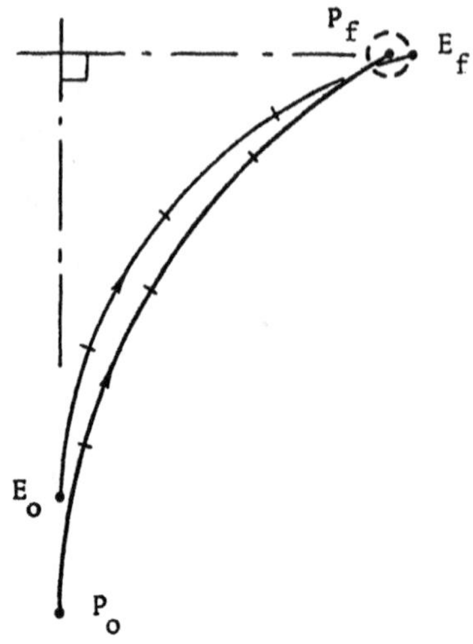

Fig. 15 Critical Trajectories in Region III

Note that as the product $\gamma\omega$, which is E's lateral acceleration, approaches 1, which is P's lateral acceleration , $\tau \to 0$ and $(r_f)_c \to 0$. This agrees with a result due to Cockayne [6], proving that $\gamma\omega \leq 1 \to (r_f)_c = 0$.

Contours of constant $(r_f)_c$ are shown in a parameter space $(\gamma, \frac{1}{\omega})$ in Fig. 16. These contours pass smoothly in and out of all 3 regions I, II, III.**

It remains to demonstrate that $(r_f)_c$ corresponds to the largest <u>closed</u> "barrier" $J(r,\Theta,H)$ = constant. Backwards construction of all the optimal paths, terminating with $r = (r_f)_c$, and with $H_f = \Theta_f$ or $H_f \neq \Theta_f$ (no singular arc), reveals that the surface formed by these paths does not reach the symmetry axis $\Theta = 0$ except at the neutral headings $H = 0$ and $H = \pi$. This is illustrated by 3 sections of this surface shown in Figs. 17 a,b,c.

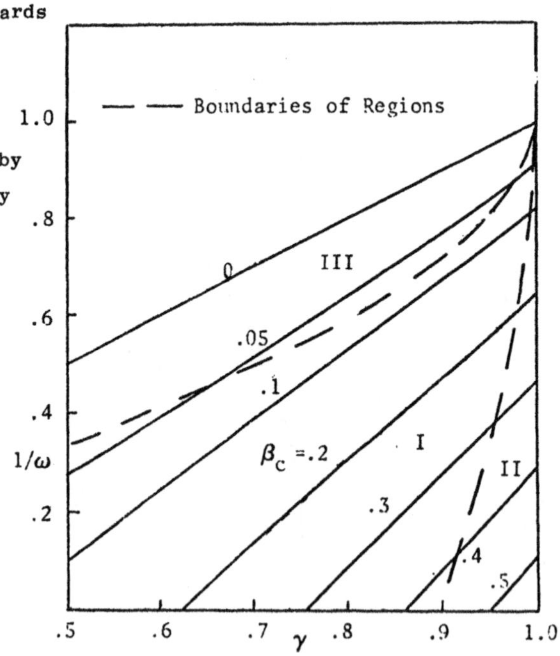

Fig. 16 Variation with Speeds and Turn-Rates of Minimum Required Capture Radius

**An entirely equivalent analysis has been carried out by S. Sharma and T. Miloh, not yet published.

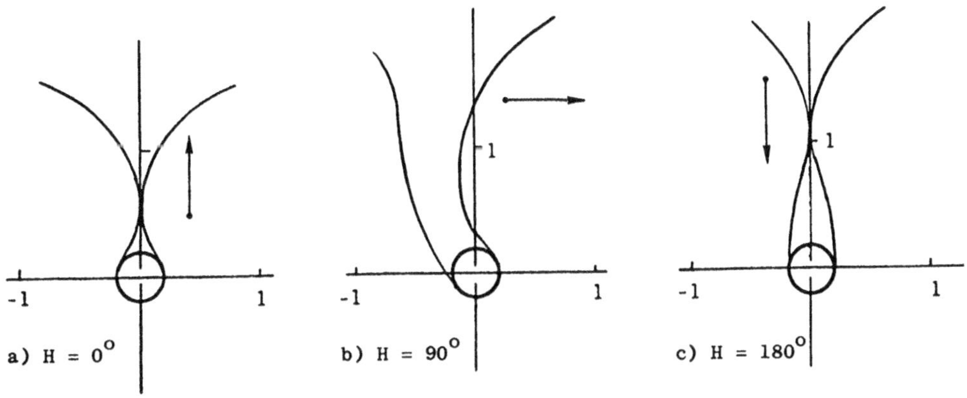

Fig. 17 Appearance of Barrier for Parameters in Region I

The barrier must be closed, at headings other than $H = 0$ and π, by finding those initial states from which E can barely return to a neutral heading $H = 0$ or π before it is "too late." Thus if E wishes to return to $H = 0$, P will reduce θ to zero and E must reach a heading $H = 0$ before r falls below $\gamma - \frac{\Upsilon}{\omega}$. This constitutes a preliminary game terminating at $H = 0$, $\theta = 0$, for which the terminal gradient (J_r, J_θ, J_H) is no longer unique, and the critical trajectories end at $r = \gamma - \frac{\Upsilon}{\omega}$. The paths of this preliminary game may be constructed backwards and P's strategy can be quite complex (see ref. [7]). After reaching the <u>decision point</u> $(\gamma - \frac{\Upsilon}{\omega}, 0, 0)$ E then decides whether to break right or left. A similar situation arises if E wishes to return to the "head-on" <u>decision point</u> $(\gamma + \frac{\Upsilon}{\omega}, 0, \pi)$. Naturally E chooses the more accessible of the two decision points, and from points on the critical barrier he can barely reach one of them. Typical composite barrier sections are shown in Figs. 18 a,b,c. The edges of the

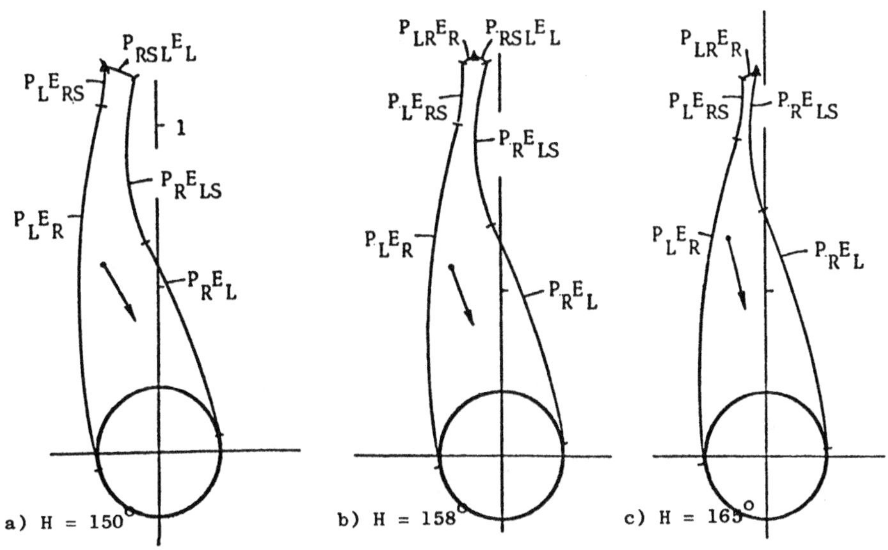

Figure 18.

of the composite barrier are either dispersive (marked ▲ in Figs. 18) or constitute the paths themselves, so that Bernhard's conditions are met [5].

A more complicated barrier in a game with identical equations of motion occurs in the "Role Determination" problem in which one airplane can destroy another if the latter comes directly in front of the former. A composite barrier separates states (r,Θ,H) leading to victory by one or the other. Space does not permit further discussion of this interesting problem, but it is described in ref. [8].

Example 5 The "Deadline Game" ([1], p. 265) and related problems.

For what initial positions of pursuer P (speed 1) and evader E (speed $w > 1$) can E pass between P and the x-axis (see Fig. 19), moving in the positive x-direction without coming closer to P than a specified distance ℓ ? The answer to this "game of kind" is obviously obtainable if we can solve the following equivalent "game of terminal payoff:" What is the distance ℓ of closest approach from any given starting position, assuming that P minimizes ℓ and E maximizes ℓ while passing between E and the x-axis?

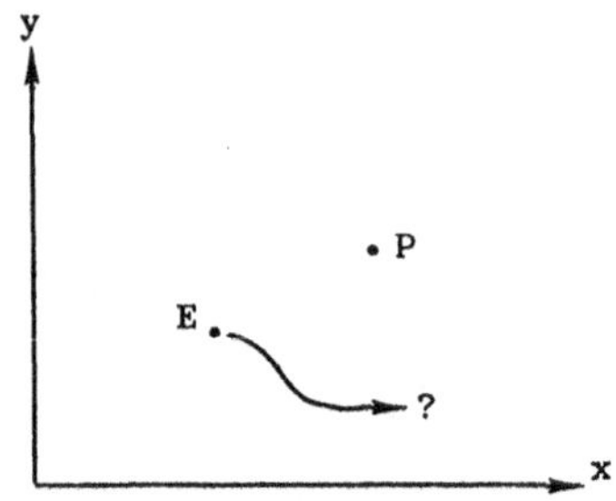

Figure 19

The positions corresponding to any particular value of ℓ will lie on a surface in the 3-dimensional state-space (x,y_P,y_E), where x denotes x_P-x_E, and the limiting surface as $\ell \to 0$ can be obtained by purely geometrical reasoning: $\ell > 0$ provided that the Apollonius circle (of positions which E can reach before P) does not reach the x-axis. As $\ell \to 0$ this circle must tend to tangency with the x-axis (Fig. 20); i.e., $y_c \to \sqrt{(EC)(PC)}$. But $y_c = y_P + \frac{y_P - y_E}{w^2-1}$ and $\frac{EC}{w^2} = \frac{PC}{1} = \frac{EP}{w^2-1}$. Hence, in the limit, $w^2 y_P - y_E = w(EP)$,

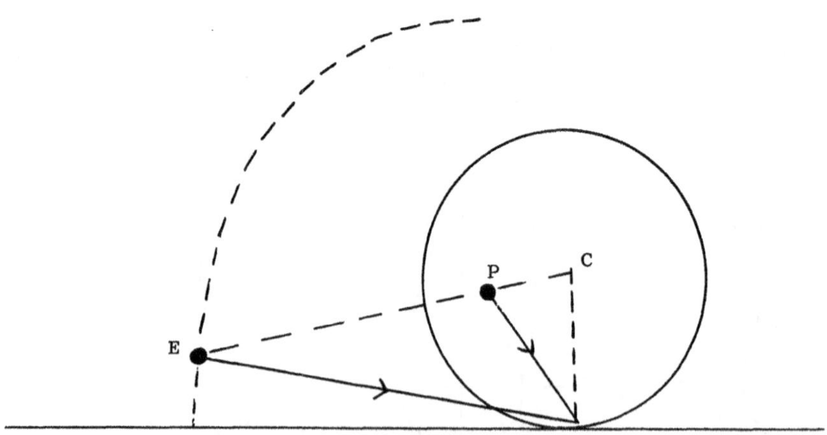

Figure 20

which defines an ellipse, eccentricity $1/w$, focus at P and minor axis along the x-axis (dotted in Fig. 20).

For $\ell > 0$, Isaacs analyzes the optimal paths in this game and concludes that they have two phases: a first, straight-line phase during which the distance PE is decreasing, and a second, curved, phase during which the distance PE remains equal to ℓ. The second phase terminates with E's path tangent to the x-axis at E_f and P, now at P_f, moving directly towards E_f (see Fig. 21) after which the distance PE increases. P's curved path is determined by the maximization of $-\frac{dy_E}{d\theta}$ subject to $PE = \ell$, which determines, in turn, E's curved path. The first phase straight paths are tangent to the second phase curved paths.

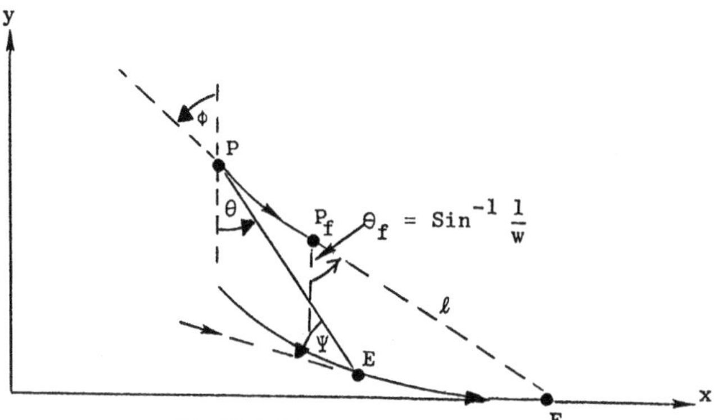

Figure 21

To verify these features, let $J(\vec{r}_E, \vec{r}_P)$ denote the closest approach distance ℓ corresponding to initial positions $\vec{r}_E$ and $\vec{r}_P$ of E and P, respectively. If $\hat{\beta}_E$ and $\hat{\beta}_P$ denote unit-vectors parallel to the velocities of E and P respectively, the main equation is:

$$0 = \underset{\hat{\beta}_E}{\text{Max}}\ \underset{\hat{\beta}_P}{\text{Min}}\ (J_{\vec{r}_E} \cdot w\hat{\beta}_E + J_{\vec{r}_P} \cdot \hat{\beta}_P)$$

$$= w\left|J_{\vec{r}_E}\right| - \left|J_{\vec{r}_P}\right| ,$$

with $\hat{\beta}_E$, $\hat{\beta}_P$ parallel to $J_{\vec{r}_E}$, $-J_{\vec{r}_P}$ respectively.

The gradient vectors $J_{\vec{r}_E}$, $J_{\vec{r}_P}$ are, furthermore, constant along unconstrained paths, implying straight-line motion. Moreover, $J(\vec{r}_E, \vec{r}_P)$ is a function $J(x, y_P, y_E)$, where $x = x_P - x_E$, so that $J_{x_E} = -J_{x_P}$. It now follows that the "controls" Ψ, ϕ (see Fig. 21) satisfy:

$$\frac{w}{\sin \Psi} = \frac{1}{\sin \phi} \tag{A}$$

During the final phase, however, E must observe the "state constraint" that EP not be allowed to decrease further. This requires that E's control Ψ be a function $\tilde{\Psi}(\phi,\Theta)$ defined by

$$w\cos(\Psi-\Theta) = \cos(\phi-\Theta) \qquad \text{(B)}$$

[If an objection is raised to this implicit assumption that E knows P's present control ϕ, E can achieve as small a change in EP as he pleases by utilizing knowledge of P's control in the recent past, as close as necessary to the present.]

The constrained main equation, using coordinates r,Θ,y_E in place of x,y_P,y_E where $r = EP$, is:

$$\underset{\substack{\phi \\ \Psi=\tilde{\Psi}(\phi,\Theta)}}{\text{Min}} \left(J_\Theta \dot{\Theta}(\Theta,\phi,\tilde{\Psi}) + J_{y_E} \dot{y}_E(\Theta,\phi,\tilde{\Psi})\right) = 0 \quad ,$$

which implies the stationarity of $\frac{dy_E}{d\Theta}$ w.r.t. ϕ. [Obviously, $-\frac{dy_E}{d\Theta}$ should be maximized.] It is also easily verified, by comparing constrained and unconstrained main equations in coordinates r,Θ,y_E, that Ψ and ϕ must be continuous at junctions of the unconstrained and constrained paths. This proves that the unconstrained paths are straight tangents to the second phase curved paths. The curved path directions Ψ,ϕ are determined by (A) and (B) (see [9]) and it is shown in [1] equation (9.5.9) that the resulting paths, for given ℓ, are given by

$$\left.\begin{aligned}
y_E &= \frac{\ell}{w^2-1}(kw - w\Theta - w^2\cos\Theta) \\
y_P &= \frac{\ell}{w^2-1}(kw - w\Theta - \cos\Theta) \\
x_E &= \frac{\ell}{w^2-1}(w^2\sin\Theta - w) + x_{E_f} \\
x_P &= \frac{\ell}{w^2-1}(\sin\Theta - w) + x_{E_f} \quad ,
\end{aligned}\right\}$$

where $k = \sqrt{w^2-1} + \text{Sin}^{-1}\frac{1}{w}$.

These paths may be constructed geometrically by rolling the Apollonius circle, now of radius $w\ell/(w^2-1)$, along a line through E_f parallel to the y-axis (see Fig. 22). Indeed, the instantaneous center of rotation of the frame, rigidly attached to E and P during the curved phases, must lie on this Apollonius circle. P, wishing to maximize $-\frac{dy_E}{d\Theta}$ where Θ measures the orientation of this frame, chooses the instantaneous center as far to the right as possible in Fig. 22. Hence the above geometrical construction.

The inclusion of the first phase straight tangents gives $y_P(\Theta,\tau)=y_P(\Theta)-\tau y_P'(\Theta)$, etc., as in [1], equation (9.5.10). This expresses $y_E, y_P, x_E-x_{E_f}, x_P-x_{P_f}$ in terms of ℓ,Θ,τ:

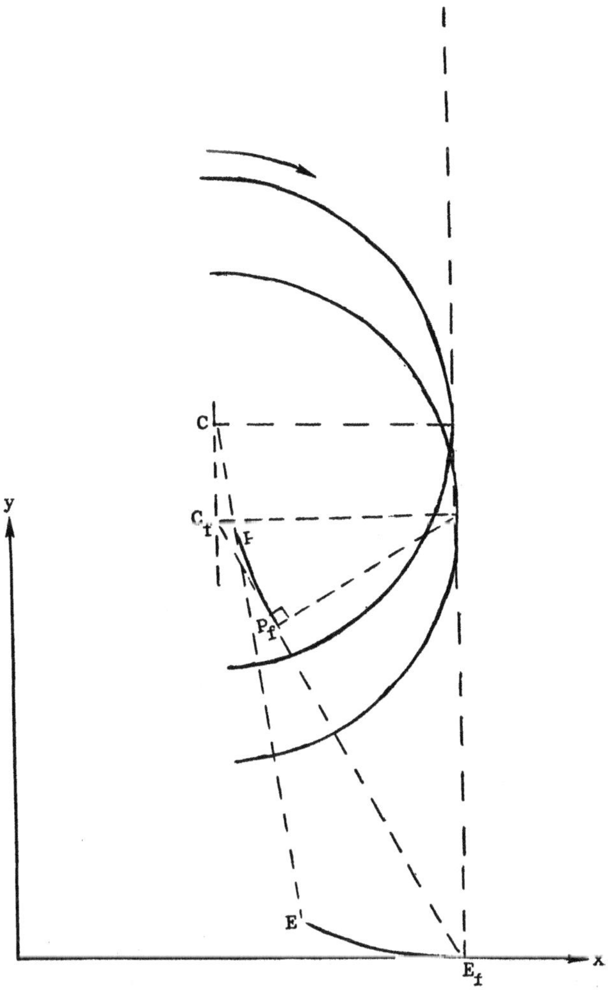

Figure 22.

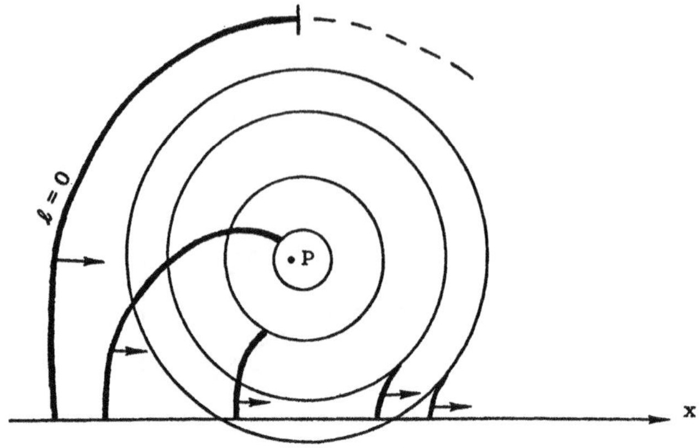

Figure 23.

$$\left.\begin{aligned} y_E &= \frac{\ell}{w^2-1}\left\{kw - w\theta - w^2\cos\theta + w\tau(1 - w\sin\theta)\right\} \\ y_P &= \frac{\ell}{w^2-1}\left\{kw - w\theta - \cos\theta + \tau(w - \sin\theta)\right\} \\ x_E &= \frac{\ell}{w^2-1}\left\{w^2\sin\theta - w - w^2\tau\cos\theta\right\} + x_{E_f} \\ x_P &= \frac{\ell}{w^2-1}\left\{\sin\theta - w - \tau\cos\theta\right\} + x_{E_f} \end{aligned}\right\}$$

A section, for fixed y_P, of the surfaces corresponding to various ℓ, is sketched in Fig. 23, E's optimal direction, indicated by arrows, being perpendicular to the local ℓ-contour.

A somewhat similar problem arises if E wishes to pass between two slower pursuers. Again the final phase is one of curved motion, in this case by E and the closer pursuer, and the state-space is again essentially 3-dimensional. Space does permit further elaboration, but the complete solution appears in reference [10]. Appendix A lists certain errata in reference [10].

Returning to the deadline game, suppose that E is permitted to escape in either direction along the x-axis. This is the two-sided deadline game, [1], p. 260. Figure 24 shows a section of the surfaces S_1, S_2 corresponding to a fixed y_P and a single value of ℓ, and to escape to the right, left respectively. The surfaces S_1, S_2 in the 3-dimensional state-space (x, y_P, y_E) intersect along a "dispersive edge," designated by Ⓓ in Fig. 24, with the following property: if E at Ⓓ chooses, for example, to escape to the right, he will remain on S_1 but will move off of surface S_2 to the "capture" side, i.e., he will no longer be in a position to escape to the left.

The surfaces S_1, S_2, together with their junction Ⓓ, form a composite semi-permeable surface (see also [5]), enclosing a region from which E cannot

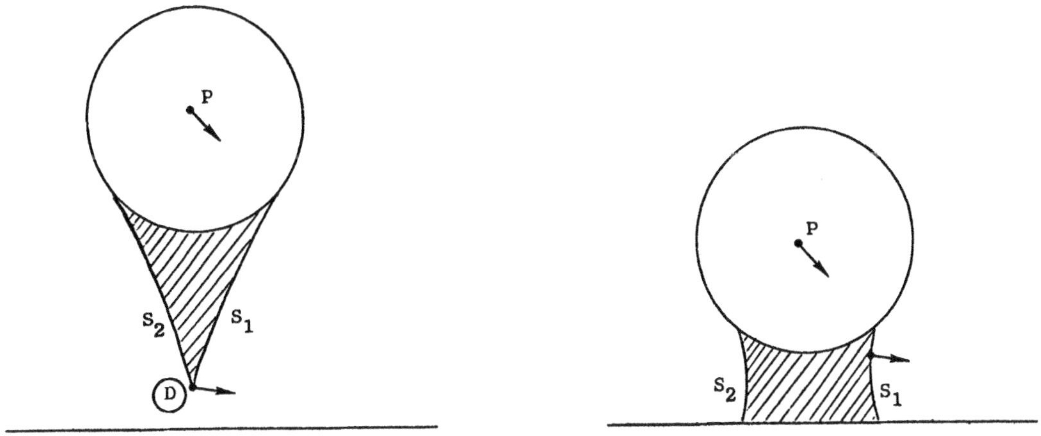

Figure 24.

escape (without coming closer to P than ℓ). Analytically this is expressed by $J^{(2)}_{\vec{r}_E} \cdot w\hat{\beta}^{(1)}_E + J^{(2)}_{\vec{r}_P} \cdot \hat{\beta}^{(1)}_P < 0$ at (D), which, because of the main equation, may be expressed in the symmetric form:

$$\hat{\beta}^{(1)}_E \cdot \hat{\beta}^{(2)}_E - \hat{\beta}^{(1)}_P \cdot \hat{\beta}^{(2)}_P < 0 \quad \text{at (D)} .$$

Patrolling a Channel

An interesting extension of the one-sided deadline game is the game: "patrolling a channel," also discussed in [1]. Here E can choose whether to pass along the channel (to the right in Fig. 25) above or below P. If the width L of the channel is less than a critical width:

$$L_c = \frac{2\ell w}{w^2-1}\left(\frac{\pi}{2} + k\right) ,$$

the surfaces S_1 and S_2, corresponding to passage below or above P respectively, form with the ℓ-circle around P the boundary of a capture region (shaded).

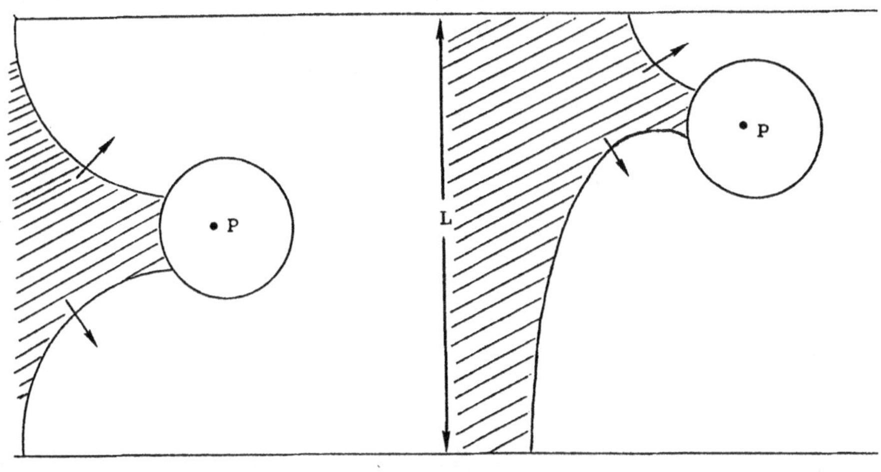

Figure 25.

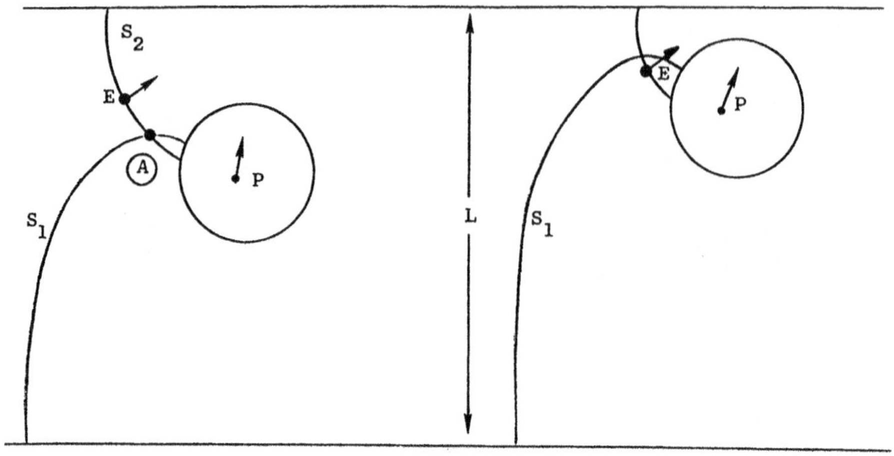

Figure 26.

If, however, $L > L_c$, the surfaces S_1 and S_2 intersect outside the ℓ-circle in an "attractive edge," designated by A (see Fig. 26), which is such that if, for example, E chooses to pass above P, E remains on S_2 but moves off of S_1 to the escape side, i.e., $\hat{\beta}_E^{(1)} \cdot \hat{\beta}_E^{(2)} - \hat{\beta}_P^{(1)} \cdot \hat{\beta}_P^{(2)} > 0$ at (A). The surfaces S_1 and S_2 thus clearly fail to form a composite semi-permeable surface -- if P guards against passage above P, E can pass below P.

The Cornered Rat

A further extension is the "cornered rat" game, also mentioned by Isaacs in [1]. The state-space is now essentially 4-dimensional and the locus of E's positions for fixed P, corresponding to given ℓ and to escape along the x or y directions, is sketched in Figs. 27 a,b,c. For P sufficiently close to the corner, as in Fig. 27a, the hypersurfaces S_1 and S_2 do not intersect. For P somewhat further from the corner they intersect in two "dispersal hyperedges," denoted by (D) in Fig. 27b. But for P sufficiently far from the corner, as in Fig. 27c, S_1 and S_2 intersect in an "attractive hyperedge" (A). The hypersurfaces thus again fail to combine into a composite semi-permeable hypersurface.

The critical positions P^* and E^*, corresponding to a change from dispersive

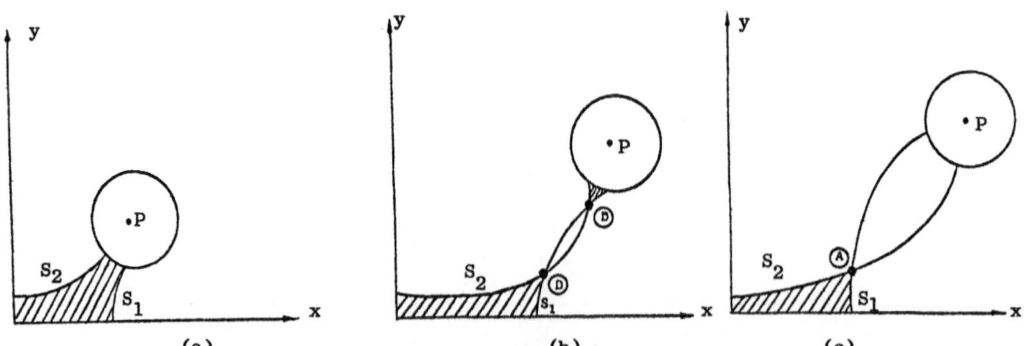

Figure 27.

hyperedge (D) to attractive hyperedge (A) , satisfy:

$$\hat{\beta}_E^{(1)} \cdot \hat{\beta}_E^{(2)} - \hat{\beta}_P^{(1)} \cdot \hat{\beta}_P^{(2)} = 0 .$$

It is straightforward to verify that this implies: $\tau_1 = \tau_2 = 1$ and $\theta_2 = -\theta_1$ (θ_2 being measured from the x-axis rather than the y-axis), and that the P^*- and E^*-loci, sketched in Fig. 28, are given by:

$$\left.\begin{aligned}
\frac{w^2-1}{\ell} x_P^* &= kw + w\theta_1 - \cos\theta_1 + w + \sin\theta_1 \\
\frac{w^2-1}{\ell} y_P^* &= kw - w\theta_1 - \cos\theta_1 + w - \sin\theta_1 \\
\frac{w^2-1}{\ell} x_E^* &= kw + w\theta_1 - w^2\cos\theta_1 + w + w^2\sin\theta_1 \\
\frac{w^2-1}{\ell} y_E^* &= kw - w\theta_1 - w^2\cos\theta_1 + w - w^2\sin\theta_1
\end{aligned}\right\}$$

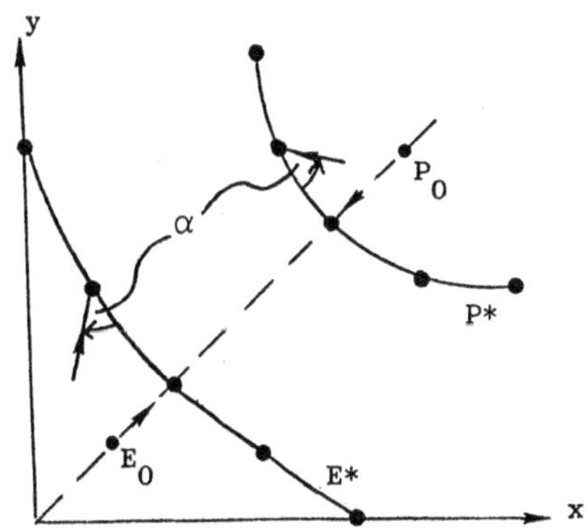

Figure 28.

It may further be easily verified that $P^*E^* = \ell\sqrt{2}$ and the arc-lengths of these loci satisfy $|d\vec{r}_E^*| = w|d\vec{r}_P^*|$. We anticipate that for positions P outside of the P^*-locus there is a locus of positions E inside the E^*-locus from which P and E move in straight lines towards a decision surface consisting of the P^*- and E^*-loci, for various ℓ, at which time E chooses which side of P to pass. If so, the directions during the "delayed option" phase of the game are determined by:

$$0 = \frac{dV^*}{d\theta_1} = J^{(3)}_{\vec{r}_E} \cdot \vec{r}_E^{*\prime}(\theta_1) + J^{(3)}_{\vec{r}_P} \cdot \vec{r}_P^{*\prime}(\theta_1) ,$$

which may be rewritten:

$$\hat{\beta}_E^{(3)} \cdot \vec{r}_E^{*\prime}(\theta_1) - w\hat{\beta}_P^{(3)} \cdot \vec{r}_P^{*\prime}(\theta_1) = 0 ,$$

the superscript (3) denoting the delayed-option phase. Together with the arc-length relation this implies that the delayed-option phase straight paths make equal angles, say α, with the E^*- and P^*-loci, as indicated in Fig. 28. In particular, for the symmetric positions E_0 and P_0 the two players move directly toward each other before E decides which way to pass. The locus of positions E corresponding to a delayed option strategy, when P is outside the P^*-locus, can now be computed. Figure 29 shows a sketch of how this locus should look. Note that the last equation on page 24 is satisfied automatically by $(\hat{\beta}_E^{(3)},\hat{\beta}_P^{(3)}) = (\hat{\beta}_E^{(i)},\hat{\beta}_P^{(i)})$, $i = 1$ or 2, since $(\vec{r}_E^{\,*}(\theta_1), \vec{r}_P^{\,*}(\theta_1))$ belongs to both hypersurfaces S_1 and S_2 for all θ_1. The surface S_3 thus joins smoothly onto the surfaces S_1 and S_2.

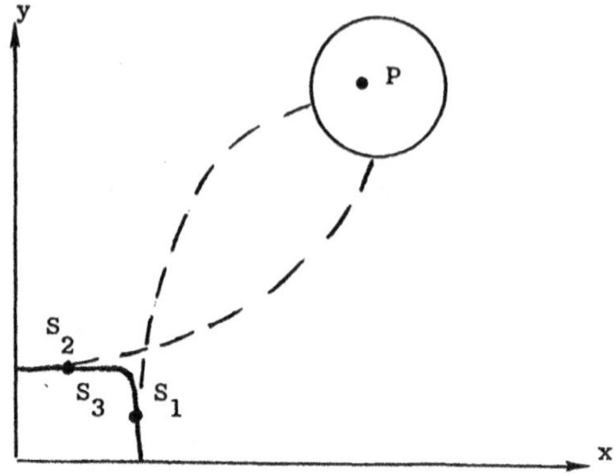

Figure 29.

Figure 29 replaces Fig. 27c and, together with Figs. 27 a,b, defines a composite barrier. Note that on the E^*- and P^*-loci the quantities $J_{\vec{r}_E}^{(i)} \cdot w\hat{\beta}_E^{(3)}(\alpha) + J_{\vec{r}_P}^{(i)} \cdot \hat{\beta}_P^{(3)}(\alpha)$, $i = 1$ and 2, which are homogeneous linear in $\sin\alpha$ and $\cos\alpha$, vanish for two distinct α's not differing by π, and therefore vanish identically. The delayed-option paths thus arrive tangentially to both hypersurfaces S_1 and S_2, thus fulfilling necessary conditions described in [5].

REFERENCES

1. Isaacs, R., Differential Games, Wiley 1966 and Krieger, 1975.
2. Flynn, J., "Lion and Man: The Boundary Constraint," SIAM J. Contr. 11 (1973), 397-411.
3. Lewin, J. and J.V. Breakwell, "The Surveillance-Evasion Game of Degree," JOTA, 16, Nos. 3/4, (Aug. 1975), 339-353.
4. Dobbie, J.M., "Solution of Some Surveillance-Evasion Problems by the Methods of Differential Games," Proc. 4th Internat'l. Conf. on Operational Research, MIT, (Wiley, 1966).
5. Bernhard, P., "Conditions de Coin pour les Jeux Differentiels," Seminaire sur les Jeux Differentiels, Centre d'Automatique, Paris (1971).
6. Cockayne, E., "Plane Pursuit with Curvature Constraints," SIAM J. Appl. Math., Vol. 15, No. 6, Nov. 1967, pp. 1511-1516.
7. Breakwell, J.V. and A.W. Merz, "Minimum Required Capture Radius in a Coplanar Model of the Aerial Combat Problem," submitted to AIAA Journal for publication.
8. Olsder, G.J. and J.V. Breakwell, "Role Determination in an Aerial Dogfight," Int. J. of Game They., Vol. 3, Issue 1, pp. 47-66, 1974.
9. Breakwell, J.V., "Pursuit of a Faster Evader," The They. & Appl. of Diff. Games, pp. 243-256, 1975.
10. Hagedorn, P. and J.V. Breakwell, "A Differential Game with Two Pursuers and One Evader," JOTA, Vol. 18, No. 1, Jan. 1976 and Multicriteria Decision Making and Differential Games, ed. George Leitmann, Plenum Press, 1976.

APPENDIX A ERRATA IN REFERENCE [10]

Page 446, eq'n. (14): $\dot{\theta} = \dot{\phi}_1 \underset{\uparrow}{+} [ws/2\sqrt{1-w^2c^2}\,\dot{\phi}_1]$

Page 449, eq'n. (27): delete the terms $-2sc\tau$

Page 449, eq'n. (29), 2nd line: $\underset{\uparrow}{+} (2c^2-1)\tau$

Page 449, eq'n. (31), left member: $y_{P_2}(\phi_1, \tau) = \ldots$

Page 451, eq'n. (33): replace each z_r by z_r^2

TWO LECTURES ON COURNOT'S PROBLEM

J. CASE, Baltimore

1. COURNOT'S PROBLEM

The problem Cournot originally considered [1] was that facing two well owners in a small village. Since they produce indistinguishable products, he observed, they must charge the same price. And presumably that price will rise or fall in such a way that all the water pumped out of the ground in a given day will be sold that day. Let q_i be the quantity of water pumped by owner i on a given day. Then it is reasonable to assume that the average price p brought by a gallon of water that day will be some decreasing function f of the total quantity q_1+q_2 of water sold. Cournot began a trend in mathematical economics by refusing to specify what function f he had in mind. But in fact little of his meaning is lost if we assume

$$(1.1) \qquad p = A - B\,(q_1 + q_2) = f(q_1 + q_2)\ ,$$

the price p being quoted in dollars per quart. Then owner i will earn roughly

$$(1.2) \qquad Aq_i - Bq_i\,(q_1 + q_2) \qquad\qquad i = 1,2$$

dollars that day. And if we substitute

$$(1.3) \qquad q_1 = Ax/B \quad \text{and} \quad q_2 = Ay/B$$

the expressions (2) become

$$(1.4) \qquad \begin{aligned} (A^2/B)\ (x - x(x+y)) &= (A^2/B)\ \ \pi_1(x,y) \\ (A^2/B)\ (y - y(x+y)) &= (A^2/B)\ \ \pi_2(x,y) \end{aligned}$$

The functions π_1 and π_2 so defined describe the reward structure of the game G between the two well owners.

The rules of G are that x and y, the players' respective actions, must be non-negative numbers. So G's salient features are all present in the representation

$$(1.5) \qquad 1 \ \underset{x \geq 0}{\text{maximize}} \ \pi_1(x,y) \ , \quad 2 \ \underset{y \geq 0}{\text{maximize}} \ \pi_2(x,y) \ .$$

Of course the imperatives "maximize" are impossible to perform without collaboration, since the players cannot each have fore-knowledge of the other's action. So "maximize" should really be replaced by "make large" in the representation (1.5). But as will be seen, approximate maximization can sometimes be achieved even though strict maximization can not. At any rate, (1.5) is a compact (if not concise) representation of G, and we shall make frequent use of such representations hereinafter.

This problem may be easily treated by Cournot's method. One calculates

$$(1.6) \qquad \frac{\partial \pi_1}{\partial x} = 1-2x-y \quad \text{and} \quad \frac{\partial \pi_2}{\partial y} = 1-x-2y \ ,$$

so that

$$(1.7) \qquad \begin{array}{l} x^*(y) = (1-y)/2 \ , \ y^*(x) = (1-x)/2 \ , \\ \text{and } \pi_1{}^*(y) = (1-y)^2/4 \ . \end{array}$$

The graphs of $x^*(y)$ and $y^*(x)$ are the straight lines shown in figure 1A, and Cournot's well known iterative process converges to their junction at $(x^*,y^*) = (1/3, 1/3)$.

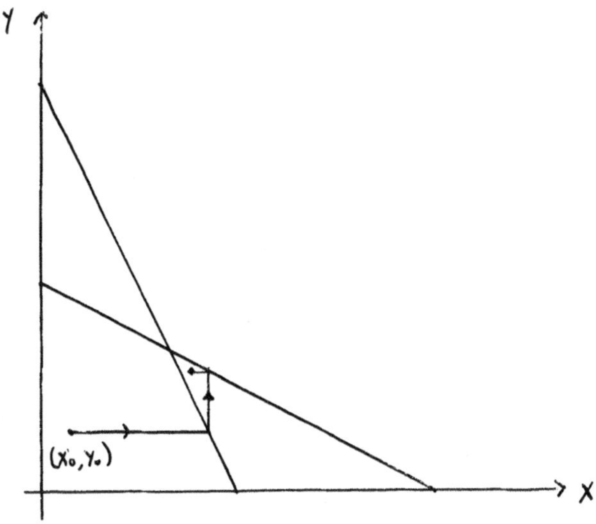

Figure 1A

But more to the point, the game may be analyzed in terms of "robust strategies."

To that end, observe that $\pi_1(x,y)$ is a linear function of y for each fixed x, since it may be written

$$(1.8) \qquad z = \pi_1(x,y) = x(1-x) - xy .$$

Its graph in the yz-plane is a straight line, which meets the y-axis at (1-x,0) and the (vertical) z-axis at (0,x(1-x)). So $x > 1$ implies $\pi_1(x,y) < 0$ for every y, and the actions $x > 1$ are all dominated by $x = 0$. And if $x > 1/2$, then $\pi_1(1-x,y) > \pi_1(x,y)$ for every y, since their graphs have the same z-intercept and the latter slopes more steeply downward than the former, as indicated in figure 1B. So every action $x > 1/2$ is dominated by some action $0 < x \leq 1/2$. But the latter actions are undominated since each of them is optimal against some y.

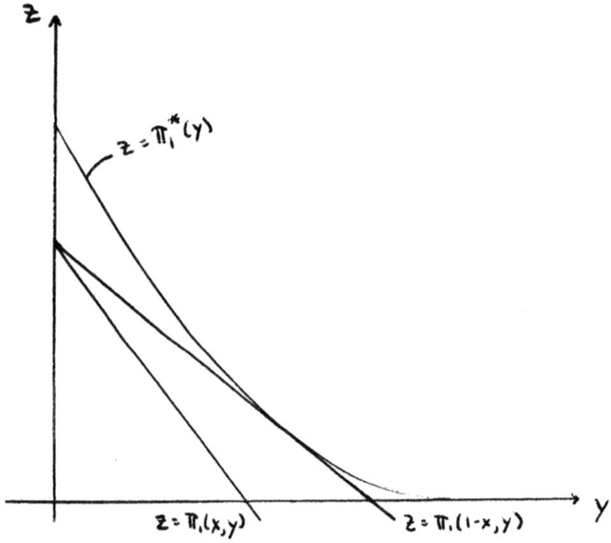

Figure 1B

Indeed the graphs of $\pi_1(x,y)$ and $\pi_1{}^*(y)$ are tangent to one another at $y = 1-2x$, and the graph of $\pi_1{}^*(y)$ is the envelope of the straight lines (1.8) for which $0 \leq x \leq 1/2$.

Clearly the straight lines do approximate their envelope over fairly large intervals in the y-axis. For instance (see Table 1C) $\pi_1(.3,y) > .75\ \pi_1{}^*(y)$ for $0 \leq y \leq .6$, and $\pi_1(.3,y) > .93\ \pi_1{}^*(y)$ for $.2 \leq y \leq .5$! Indeed because the dominance relations established earlier make it unlikely that 2 would ever take an action $y > 1/2$, $x = .3$ seems a very robust strategy in the sense that it is nearly optimal against a wide range of the anticipated y-values. On the other hand, the values given in table 1C make it clear that $x = .1$ and $x = .5$ are not particularly robust.

Table 1C

y	π_1*(y)	π_1(.1,y)	%	π_1(.3,y)	%	π_1(.5,y)	%
0	.25	.09	36	.21	84	.25	100
.05	.226	.085	38	.195	86	.225	99+
.1	.203	.08	39	.180	89	.200	99
	.181	.075	41	.165	91	.175	97
.2	.161	.07	43	.150	93	.150	93
	.141	.065	46	.135	96	.125	87
.3	.123	.06	99	.120	98	.100	81
	.106	.055	52	.105	99+	.075	71
.4	.090	.05	56	.090	99+	.050	56
	.076	.045	59	.075	99	.025	33
.5	.063	.04	63	.060	95	0	0
	.051	.035	69	.045	88	-	-
.6	.040	.03	75	.030	75	-	-
	.031	.025	81	.015	48	-	-
.7	.023	.02	87	0	0	-	-
	.016	.015	94	-	-	-	-
.8	.010	.01	99+	-	-	-	-
	.0056	.005	89	-	-	-	-
.9	.0025	0	0	-	-	-	-
	.0006	-	-	-	-	-	-
1.0	0	-	-	-	-	-	-

In playing the game, the players may feel quite certain that their opponent will draw some water, but will not know how much. So each should seek a robust action and may assume that his opponent wants one too. Each might take, for instance, the action .3. If both do, their choices will each be 98% optimal against their opponent's. Moreover their joint earnings will be .240 which is 96% of the .250 they could earn together if they became partners. So there is little incentive here to cooperate.

Of course the players could also take the actions x = 1/3 = y, which are only a little less robust than x = .3 = y, and which are the Cournot-Nash equilibrium strategies as well. But in that case they would together earn only 2/9 instead of the 1/4 they could get by cooperating. And 2/9 is only 89% of 1/4. So their quest for exact optimality instead of approximate optimality would result in an actual loss of income, as well as of robustness, in this case.

Another solution of Cournot's game has been proposed which, because of its popularity [9] over the years, richly deserves mention here. It is the so-called Stackelberg solution, and purports to be the outcome fame would have if one well-owner retained the other's assistant as a spy. For then the spying owner would always know his opponent's actions in advance, and could consistently choose his own to be optimal against them.

Let us suppose, for definiteness, that owner 2 retains an agent to spy on 1. Then 2 can arrange that

$$y = y^*(x) = (1-x)/2 \tag{1.9}$$

always. In doing so, 2 becomes the so-called follower, since his action is now chosen after (and in response to) his opponent's. Owner 1 is called the leader, and the best he can do in his "predicament" is to maximize

$$\pi_1(x, y^*(x)) = x(1-x)/2 \tag{1.10}$$

thru his choice of x. Then

$$x = 1/2 \quad \text{and} \quad y = y^*(1/2) = 1/4. \tag{1.11}$$

so

$$\pi_1 = \pi_1(1/2, 1/4) = 1/8 \quad \text{and} \quad \pi_2 = \pi_2(1/2, 1/4) = 1/16. \tag{1.12}$$

There is profit indeed in leadership!

But something is terribly wrong with this conclusion, for crime has not paid! That is, it was 2 (the follower) who introduced asymmetry into the game by his industrial espionage, and it is he who has suffered from that asymmetry. One suspects that better use can be made of the information supplied by his agent.

The sensible way to use such extraordinary intelligence is, of course, in conjunction with the other weapons at his disposal. And the most powerful of those are his robust actions. The reaction function (1.9) is too volatile for a player with robust actions in his arsenal; it allows his opponent to "tool" him about like a monkey on a string. The players of various games cannot afford to let their opponents rule them so.

While 2 is using the rule (1.9) for instance, 1 can force him to reduce his presence in the market from y = .33 to y = .25 by the substitution of x = .5 for x = .33. But if 2 elects to react in a more relaxed fashion to intelligence received by using, for instance, the less volatile reaction function

(1.13) $$y = .4 - .24x ,$$

then the substitution of x = .5 for x = .33 causes 2 to retreat only from y = .32 to y = .28. Yet his earnings against the two x-values, namely

(1.14) $$\pi_2(1/2,.28) = .0616 = 1/16 \times .9856$$
$$\pi_2(1/3,.32) = .1109... = 1/9 \times .998...$$

compare very favorably with those he could have earned using (1.9). Moreover the "hesitant follower" who uses (1.13) is not led so far astray as the one who chooses (1.9). For the maximum of

(1.15) $$\pi_1(x,.4-.24x)$$

occurs at x = .395, and not at x = .5. So a modified Stackelberg solution of x = .395, y = .305 emerges when 2 uses (1.13) and 1 maximizes π_1 subject to that constraint.

A more relaxed fashion still of reacting to intelligence is to ignore it entirely while x remains in the interval $.1 < x < .6$, and to play the robust action y = .3 steadfastly. Then 2 can still earn

(1.16) $$\pi_2(1/2, 3/10) = 6/100 = .96 \times 1/16$$

against x = .5 and provide 1 with an even greater incentive to stop trying to dominate the market. For the maximum of

(1.17) $$\pi_1(x, 3/10)$$

occurs at x = .35. In this case, a yet more modified Stackelberg solution emerges at x = .35 and y = .3, and the leader-follower relationship is virtually obliterated. Indeed we have returned nearly to the

Cournot point $x = 1/3 = y$.

The message here is simple. Because of the robustness of the actions at his disposal, 2 has no need to react violently to intelligence received concerning 1's intent. He is at liberty to "follow" changes in x with much smaller changes in y, if any. Therefore "leadership" in the extreme sense postulated by Stackelberg can never be established, and the Stackelberg solution will never be observed in any real market.

Actually entrepreneurs realize these things instinctively, and they do not capitulate immediately to invasions of their markets. They resist. And robustness is the weapon wherewith they resist. Indeed it could be argued that the robustness of a firm's market strategy is the proper measure of its "strength" in the market.

Of course, not all firms have actions to choose from as robust as those in the linear-demand version of Cournot's game. For robustness depends on the shape of the demand curve. But there will always be some degree of robustness available, for the curves $z = \pi_2(x,y)$ which arise from fixed values of y are tangent to their envelope $x = \pi_2(x,y^*(x)) = \pi_2{}^*(x)$, so that each undominated y is optimal against some x and very nearly so against a host of nearby ones. This fact will quickly become evident to any firm that endeavors to establish the sort of market leadership that Stackelberg suggests is possible.

The Stackelberg solution is an old and beloved fiction of economic theory. But it is only that--a fiction. It is based on ignorance of the most important attribute that a competitor's action can have, namely its robustness. When that attribute is properly accounted for, one sees immediately that the degree of followership that Stackelberg postulates is impossible. Instead a much more limited form of followership presents itself which leads back to the conclusion that a solution is a pair of robust actions, better likened to Cournot's conception than to Stackelberg's. Robustness, in short, seems to bring order out of chaos in Cournot's game at least.

2. MORE PLAYERS

The same game can be played among any number of well owners. Let their names be 1, ..., N, and let their actions be $x, x_2, \ldots, x_N$ respectively. Write $y = x_2 + \ldots + x_N$. Then player 1's reward is still specified by (1.8), and his objectives are (as before) to guess what y will be and to pick an x which will be roughly optimal against it.

Player 1 can no longer assume that $y < 1/2$, though of course the individual inequalities

(2.1) $\quad x_2 \leq 1/2 \quad x_3 \leq 1/2 \quad \ldots \quad x_N \leq 1/2$

still follow from the dominance relations. But he knows what y's past values have been, for he has a record of his own previous days' actions x (t) for $t = 0, -1, -2, \ldots$ and he knows what his earnings $\pi_1(t)$ were on those days. Also

(2.2) $\quad y(t) = 1 - x(t) - \pi_1(t)/x(t)$

for $t = 0, -1, -2, \ldots$, so that the sequence $\{y(t)\}_{t=0}^{-\infty}$ can be regarded as known. So if that sequence appears sufficiently regular, player 1 may hypothesize that y(1) will fall in a "confidence interval" $\alpha < y(1) < \omega$. And if that interval is not too large, he can choose a robust action x which is effective against all the y's in it.

There are no guarantees here; the promise of profit is always at the risk of loss even against a single opponent. For player 2 alone can make $y > 1$ which prevents $\pi_1 > 0$. But that is the chance each merchant must take, both in the model and in the real market place it seeks to reflect. And the present analysis in terms of robust strategies goes far in explaining the success with which just that chance is so regularly taken. If something didn't make it easy, people wouldn't be willing to make their livings at it.

It is worthy of note that Cournot's game has been played rather extensively [12] under laboratory conditions, by juniors, seniors, and graduate students in a well-known school of business administration. Only a brief summary of the experiments and results will be given.

The game the subjects played was not exactly the one described here, since production costs were present, and only integral quantities q_1, q_2 of output were encouraged by the experimenter. But in its essential details, the game was as specified above. In particular, the price law was (1.1) with $A = 120$ and $B = 3.158$. The Cournot-Nash equilibrium for this version of the game occurs at $q_1 = q_2 = 12.667$ and the symmetric Pareto optimum is $q_1 = q_2 = 9.5$. The duopoly data were produced in a pilot study by four subjects who made output decisions in two duopoly markets for twenty-five successive "weeks." The bottom line of table 2A shows the frequency distribution of the 100 quantities produced by the four subjects in these twenty-five time periods; it is a trimodal distribution.

Stern's Results

Periods	5	8	9	10	11	12	13	14	15	16	17	18	19	22	25	Number of observations in group
1-5	1	2	2	0	0	2	1	1	1	1	1	6	0	1	1	20
6-10	0	0	0	0	1	1	4	3	3	1	2	5	0	0	0	20
11-15	0	0	1	2	1	1	5	2	2	1	0	4	1	0	0	20
16-20	0	1	0	10	1	5	2	0	0	0	0	1	0	0	0	20
21-25	0	0	0	4	0	2	9	2	3	0	0	0	0	0	0	20
1-25	1	3	3	16	3	11	21	8	9	3	3	16	1	1	1	100

Table 2A

A number of things are apparent from the table. To begin with, twenty-five of the hundred quantities produced exceed fifteen units. Of these, twenty-four occurred in the first fifteen weeks and nineteen in the first ten. This suggests that an initial learning period was required for the players to discover the need to limit production. Large outputs became increasingly rare as the players gathered experience.

Of the remaining seventy-five decisions, the most popular choices were thirteen, ten, and twelve. That is about what one would expect, for the choices between $q_i = 9.5$ (the symmetric Pareto optimal and $q_i = 12.667$ (the Cournot-Nash equilibrium decision) are the most robust actions available to players of this game. So a large number of choices lying outside the interval from (say) ten thru thirteen would cast doubt on the validity of the foregoing theory. But a majority of all the decisions taken (fifty-one of the entire hundred and thirty-three of the final forty) do lie in the interval. So theory and observation are in rough accord.

Something _not_ predicted by the theory is the relative unpopularity of the choice $q_i = 11$. Because of it, the frequency distribution has two separate modes in the robust interval $10 \leq q_i \leq 13$. Perhaps the phenomenon is a real one, and could be observed outside the laboratory as well. But one suspects that the noisy environment of a real market place would tend to obscure such narrowly separated peaks and fuse them together into one. In that case it would be impossible to assert that non-cooperative (Cournot-Nash type) behavior is more common than the cooperative (Pareto optimal) kind as it was a little in the experiment. But it would still be possible to assert that robust actions predominate over all others. One may even be able to substantiate such suspicions by running the experiment again, this time replacing the constant A in the price law (1.1) with a random variable of moderate variance.

There is much to be learned from experiments of this sort. But it is beside the point to seek precise conclusions from them thru the elimination of every source of noise. For the market place is a noisy environment, and people's behavior in it is conditioned by the fact. Instead of creating laboratory environments so noiseless as to permit the testing of exact hypotheses, one should devise hypotheses so inexact as to be testable in a noisy environment. The notion of robustness arises naturally in the attempt to phrase such hypotheses.

A second set of experiments are described in [12]. In them, the number of firms per market was permitted to vary. The interesting results concern the variation of the average price prevailing in the various markets with the number of firms in them. The results are exhibited in figure 2B below.

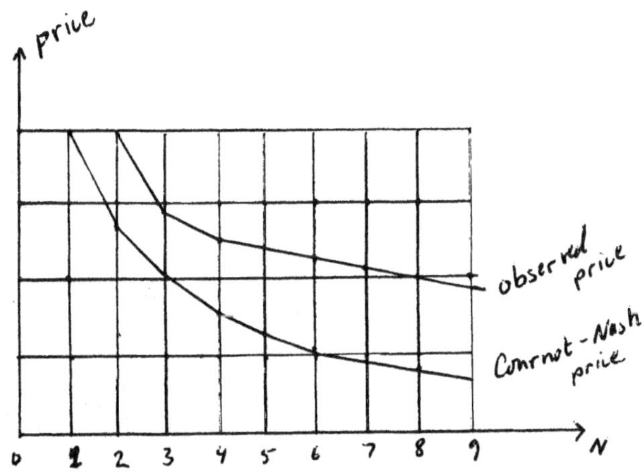

Figure 2B

The lower curve shown there represents the price that would be observed if the players really chose the Cournot-Nash equilibrium actions. And the upper one represents the average of the prices that were observed as the different sized games were repeatedly played. The fact that the observed prices exceed the predicted ones indicates that the unembellished Cournot-Nash non-cooperative theory provides an incomplete explanation at best for the players' behavior. It does, on the other hand, provide a much better explanation than the cooperative theory, according to which the prices will never fall below the monopolist's price no matter how many firms enter the market. Moreover, if one accepts the earlier conclusion that the players should ordinarily produce a little less than the equilibrium output in order to gain additional robustness at the expense of but a little additional income, it is possible to account for at least the algebraic sign of the observed discrepancy between experimental and theoretical prices on the basis of the refurbished non-cooperative theory. Less production does yield higher prices.

In short, prediction and observation are in rough agreement here both qualitatively and (to a gratifying extent) quantitatively as well. If further experimentation continues to confirm the predictions of the theory to be advanced presently, the study of games will finally have grown from a backwater of mathematics into a science in its own right. Stern's experiments are but a first step down the path of that goal.

3. SOME FORMAL DEFINITIONS

Cournot's game belongs to a rather noteworthy class of many player games. They may all be abbreviated in the form

$$(3.1) \qquad i \max_{x_i \geq 0} \quad \pi_i(x_1, \ldots, x_N),$$

since each player's action set is simply the non-negative real numbers. To play the game (3.1) each player $i = 1, 2, \ldots, N$ must choose an $x_i \geq 0$. And the reward each i receives is $\pi_i(x_1, \ldots, x_N)$. So $\pi_i, \ldots, \pi_N$ are called the reward (or payoff) functions of the game.

Definition: if each of the payoff functions π_i is unimodal in x_i for fixed x_j, $j \neq i$, and monotone decreasing in x_j for each $j \neq i$, then we call (3.1) a simply directed game.

Definition: The action x_i dominates the action x_i' if $x_i < x_i'$ and if the inequality

$$(3.2) \quad \pi_i(\ldots, x_{i-1}, x_i, x_{i+1}, \ldots) > \pi_i(\ldots, x_{i-1}, x_i', x_{i+1}, \ldots)$$

is valid for all (N-1)-tuples $(\ldots, x_{i-1}, x_{i+1}, \ldots)$ of non-negative real numbers.

The latter differs from the notion of dominance introduced in Chapter I for matrix games only in the stipulation that $x_i < x_i'$. Dominance does not have the importance for all games that it does for zero-sum ones, as will be seen from the following example in which each player i must choose between the actions $x_i = 1$ and $x_i = 2$.

(3.3)

$x_2 \backslash x_1$	1	2
1	20	10
2	30	1

$x_2 \backslash x_1$	1	2
1	5	1
2	3	500

Here the action $x_1 = 1$ dominates $x_1' = 2$. But it is easy to imagine the choice $x_1 = 2$ anyway. For 2 could offer 1 \$100 to sign a contract (iron-clad, of course) promising to choose $x_1 = 2$. So one cannot ignore the possibility that $x_1 = 2$ in this game simply because $x_1 = 1$ dominates it.

In simply directed games however, the notion of dominance (as defined above) does retain the import it has for matrix games. For clearly player i cannot increase his own payoff π_i by substituting the dominated x_i' for x_i. Nor can he expect to be bribed by an opponent or group of opponents for doing so, since all their payoffs are also decreased thereby!

It should be remarked at this time that the game (3.1) can be played, as can Cournot's game, by any number of players. Therefore the functions $\pi_1, \ldots, \pi_N$ can have any number of arguments, and equations like

$$\pi_1(x, x_2 + \ldots + x_N) = \pi_1(x, x_2, \ldots, x_N) , \tag{3.4}$$

wherein a function of two variables is equated to one of N variables, are natural. Of course (3.4) is a blatant abuse of established mathematical notation, but to us it seems easier to perpetuate that abuse than to introduce the additional notation required to correct it. We trust that our meaning will not be irretrievably lost in this way.

Actually the equation (3.4) is a special case of another, which is frequently encountered.

Definition: Let $\Omega(\cdot,\ldots,\cdot)$ be a function of N-1 variables such that the equation

$$\pi_1(x_1, x_2,\ldots,x_N) = \pi_1(x_1, \Omega(x_2,\ldots,x_N)) \tag{3.5}$$

is valid identically for non-negative $x_1,\ldots,x_N$. Then the game (3.1) is called reducible and the function Ω is called an index of opposition for it.

Cournot's game is of course reducible by the index $\Omega(x_2,\ldots,x_N) = x_2 +\ldots+ x_N$. Indeed many games are reducible and others are approximately so. The principle fruit of reducibility is that it permits us to interpret two-dimensional figures like 1B in the context of N-player games.

To appreciate the import of the definitions, a preliminary example is helpful. So let us consider the simply directed three-player game

$$(G_3) \quad i \max_{x_i} \pi_i(x_1,x_2,x_3) = (x_i/(x_1+x_2+x_3)) - x_i$$

in which the allowable values of x_i are 0, .25, and .5, i = 1,2,3. Since (G_3) is obviously symmetric in the three players and has an index of opposition Ω, it is possible to represent the payoff functions in the following tabular (matrix) form

x_i \ Ω	0	.25	.5	.75	1
0	0	0	0	0	0
.25	$.75 = \frac{3}{4}$	$.25 = \frac{1}{4}$	$.08 = \frac{1}{12}$	0	$-.05 = \frac{.1}{20}$
.5	$.5 = \frac{1}{2}$	$.17 = \frac{1}{6}$	0	$-.10 = \frac{-1}{10}$	$-.17 = \frac{-1}{6}$

It is clear therefrom that $x_i = .25$ dominates $x_i = .5$, so no player has reason to choose $x_i = .5$. Thus none need fear opposition $\Omega > .5$, and all may safely choose $x_i = .25$, which is optimal against all $\Omega \leq .5$. In that case, each player plays $x_i = .25$ against opposition $\Omega = .5$ and earn $\pi_i = .08$. Indeed, since the alternatives are only to play $x_i = .5$ and earn less or to play $x_i = 0$ and earn nothing, it seems natural to call $x_i = .25$ an <u>optimal action</u> for player i, the triple (.25, .25, .25) a <u>solution</u> of G_3, and the game G_3 itself solvable.

To make these notions precise, further definitions are required.

<u>Definition</u>: In a simply directed and reducible game with index Ω of opposition, a value Ω_o of Ω is called dominated if the inequality $\Omega(\ldots, x_{i-1}, x_{i+1},\ldots) > \Omega_o$ implies that one of the actions $\ldots, x_{i-1}, x_{i+1}\ldots$ is dominated.

<u>Definition</u>: An action x_i^* is called <u>optimal</u> in such a game if it is optimal for i against every undominated value of the index Ω of opposition.

<u>Definition</u>: An N-tuple $(x_1^*, \ldots, x_N^*)$ of optimal actions is called a <u>solution</u> of such a game, and the game itself is

called <u>solvable</u> if it possesses such a solution.

G_3 is of course a solvable game according to the above definitions. But almost all other games are not solvable in the same sense. So our definitions are overly restrictive. On the other hand, Cournot's two-player game is almost solvable, since near-optimal actions exist for both players in each of them.

<u>Definition</u>: An action x_i^0 is called robust if it is near-optimal for i against every undominated value of the index Ω of opposition.

<u>Definition</u>: An N-tuple $(x_1^0, \ldots, x_N^0)$ of robust actions is called an <u>approximate solution</u> of a game, and the game itself is called <u>approximately solvable</u> if it possesses an approximate solution.

Whether or not a particular game possesses an approximate solution often depends on the interpretation given the phrase "near-optimal" appearing in the definition of robustness. A major objective in my recent writing has been to demonstrate that many interesting and important commercial situations are properly modeled as simply directed and reducible games which are approximately solvable.

Any solution, as defined above, is a Cournot-Nash equilibrium point. And approximate solutions are in approximate Cournot-Nash equilibrium. Thus our approach is in the tradition of non-cooperative game theory. But it differs from the usual formulation of that theory in many particulars.

The non-cooperative nature of the present solution concept is, readily apparent in the game G_3, wherein the players earn 1/12 when they play "optimally." For 1 and 2 could each agree to pay 3 the sum 1/12 to choose $x_3 = 0$, thereby doubling everyone's reward. So if 3 is adequately assured that his opponents will each pay him 1/12 after he has chosen $x_3 = 0$, he should surely do so. It is only when he is forced to choose <u>without</u> such assurance that we claim $x_3 = 1/4$ to be his optimal choice.

4. <u>A Dynamic Formulation of Cournot's Game</u>. An interesting version of Cournot's game has recently been suggested by M. Simaan and T. Takayama [1]. Their idea was to introduce a dynamic price law of the form.

$$(4.1) \qquad \dot{x}(t) = -k[x(t) - (1-u_1(t) - \ldots - u_N(t))] ,$$

where $x(t)$ denotes price at time t, and $u_i(t)$ is firm l's current output level. For then the price law $x = 1-u_1-\ldots-u_N$ need not hold identically at each instant, but tends to be restored during periods wherein output levels remain constant. It may be observed that the first-order model (4.1) can never give the <u>oscillations</u> about an equilibrium price that one might expect, so that perhaps a second order model is more appropriate really. But we shall confine our attention to (4.1) for simplicity.

If each firm i has a cost-of-production function of the form

$$(4.2) \qquad K_i(u_i) = c_i u_i + (a_i/2)u_i^2 ,$$

one is led to consider the differential game

$$(4.3) \qquad \underset{u_i \geq 0}{i \text{ maximize}}\ J_i = \int_0^\infty e^{-\rho t}[(x(t)-c_i)u_i(t) - (a_i/2)u_i^2(t)]dt$$
$$\text{subject to } \dot{x} = 1-x(t) - u_1(t)-\ldots-u_N(t) ,$$

wherein the relaxation constant k has been normalized to one by choice of an appropriate time unit. The Hamilton-Jacobi method, if successful, may be expected to yield an N-tuple of feedback strategies $u_1^*(x),\ldots,u_N^*(x)$ whereby each player may observe the current price and adjust his output rate accordingly. In practice, a moving average of recent prices would probably be used instead, but we ignore such details here.

The ith player's Hamiltonian function is

$$(4.4) \qquad H_i = (x-c_i)u_i - (a_i/2)u_i^2 + p_i(1-x-u_1-\ldots-u_N) ,$$

and setting $\partial H_i/\partial u_i = 0$ yields

(4.5) $$x - c_i - a_i u_i - p_i = 0 .$$

So solving (4.5) for $u_i = u_i(x,p_i)$ and writing $V_i'(x) = p_i$ permits one to write the Hamilton-Jacobi differential equations for the unknown functions $V_1(x),\ldots,V_N(x)$. However it seems simpler, in the present case, to eliminate p_i from (4.4) by use of (4.5) and thus to obtain equations for the unknown control laws $u_1^*(x),\ldots,u_n^*(x)$ directly. When that is done, one obtains

(4.6) $$\begin{aligned} H_i &= \frac{a_i}{2} u_i^2 + a_i u_i \Omega_i - a_i(1-x)u_i - (x-c_i)\Omega_i + (x-c_i)(1-x) \\ &= \rho(V_0 + x^2/2 - x_0^2/2 - cx + cx_0 - a_i \textstyle\int_{x_0}^{x} u_i(Z)dZ) = \rho V_i(x) , \end{aligned}$$

where

(4.7) $$\Omega_i(x) = \sum_{\substack{j=1 \\ j\neq i}} u_j(x)$$

In the purely symmetric case $a_1 = \ldots = a_N = a$ and $c_1 = \ldots = c_N = c$, it makes sense to look for a symmetric solution $u_1^*(x) = \ldots = u_N^*(x) = u(x)$, wherefore $\Omega_i(x) = (N-1)u(x)$. Then (4.6) reduces to

(4.8) $$\Psi(x,u) = u^2 - P_1(x)u + Q_1(x) = \rho\left[R_1(x) - \frac{a}{a(N-\frac{1}{2})}\int_{x_0}^{x} u(Z)dZ\right] ,$$

where

(4.9) $$\begin{aligned} P_1(x) &= (a(1-x) + (N-1)(x-c))/a(N-\tfrac{1}{2}) \\ Q_1(x) &= ((x-c)(1-x))/a(N-\tfrac{1}{2}) \\ R_1(x) &= (V_0 + x^2/2 - x_o^2/2 - cx + cx_o)/a(N-\tfrac{1}{2}) \end{aligned}$$

Equation (4.8) is an integral form of the Hamilton-Jocobi equation. Differentiating yields a differential equation for the unknown strategy $u(x,\rho)$; Ψ is an integral thereof when $\rho = 0$. So setting $\Psi(x,u)$ equal to the arbitrary constant θ yields the solution

(4.10) $$u(x;\rho,\theta) = P_1(x)/2 \pm \sqrt{P_1^2(x)/4 - Q_1(x) + \theta}$$

of (4.8) when $\rho = 0$. Clearly $u(x;0,\theta)$ approximates $u(x;\rho,\theta)$ for small values if ρ, since ρ appears analytically in the differential form of (4.8).

It remains to discover the value of the constant of integration θ appearing in (4.10) and/or to find appropriate initial conditions wherefrom to integrate (4.8) when $\rho > o$. To that end, we recall the previous lecture wherein it was shown that a very good way to play Cournot's game is to ignore observed price fluctuations, and to adhere steadfastly to some (constant) robust action. Indeed such behavior seems so eminently sensible that players who do choose to use their knowledge of x should expect that some at least of their opponents will choose not to. Therefore such players should select strategies $u_i(x)$ that are effective both against opponents who do observe x and against opponents who do not.

To discover such strategies, reflect that virtually any N-tuple of stationary strategies $u_1(x), \ldots, u_N(x)$ the players might choose will in time drive the quantity $x(t)$ to a position x_∞ wherein

$$(4.11) \qquad 1 - x_\infty - u_1(x_\infty) - \ldots - u_N(x_\infty) = \dot{x}_\infty = 0 \ .$$

And once arrived in such a position, player i will have cause to be satisfied with his choice $u_i(\cdot)$ if $u_i(x_\infty)$ is roughly optimal against $\sum_{j\neq i} u_j(x_\infty)$ in the static game

$$(4.12) \qquad \max_{u_i \geq 0} \pi_i = u_i(x_\infty)[1-C_i-u_1(x_\infty)-\ldots-u_N(x_\infty)-(a_i/2)u_i(x_\infty)]$$

and dissatisfied if not. Moreover, since (4.12) is almost exactly Cournot's game, it is easy to see that $u_i(x_\infty)$ should be a robust action for that static game. We shall therefore take $u_i(x_\infty)$ to be exactly the Nash equilibrium action for (4.12), and depend upon the variation of u_i with x to provide the additional robustness one would otherwise be forced to obtain from some action slightly more conservative than the Nash equilibrium one.

In the case $N = 2$, $u = 1$, $c = 1/2$, for instance, it turns out that $u_1{}^* = u_2{}^* = 1/8$ is the static Nash equilibrium. So $x_\infty = 3/4$ from (4.11), and consequently $P_1 = 1/3$, $Q_1 = -(2x^2 - 3x + 1)/3$. Accordingly, the negative sign must prevail in (4.10) if the latter is to hold when $x = x_\infty$, and θ must be 1/64. The graph of the resulting function $u^*(x)$ appears in figure 4.2; $u^*(x)$ is positive only in the interval $.575 < x < .975$. So the solution of (4.8) fails to provide equilibrium strategies outside that interval, negative production being meaningless.

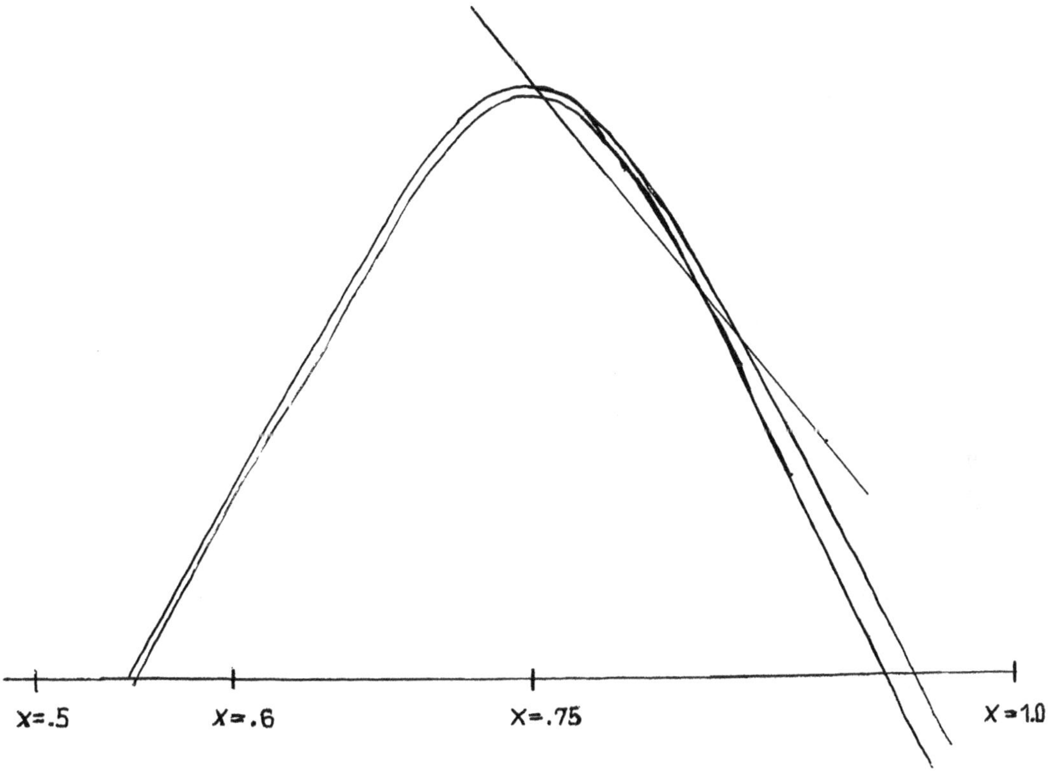

Figure 8.2

In fact another difficulty arises since, as is obvious from the portion of the line $u = \frac{1}{2}(1-x)$ shown in figure 4.2, the quantity $1-x-2u^*(x)$ changes sign not once but twice (at $x^* = .75$ and again at $x = .85$) in the interval $0 \leq x \leq 1$. So the strategies $u_1=u_2=u^*(x)$ drive $x(t)$ to x^* only from starting points $x_0 < .85$ for $x_0 > .85$, $x(t) \rightarrow 1$, which is not the result of any pair of good strategies. In short, our method has not led us to a pair of equilibrium strategies defined on the entire interval $0 \leq x \leq 1$; $u_1 = u_2 = u^*(x)$ is only in equilibrium on $0 \le x \le .85$.

It may be argued, however, that because not all prices $0 \le x \le 1$ need really be anticipated, the information provided by (4.8) does permit the construction of good strategies. The fact that the maximal joint profit in the static version of the game is achieved when $u_1 = u_2 = \frac{1}{10}$, for instance, leads one to expect $u_1 + u_2 \ge \frac{1}{5}$ whenever prevailing prices are high, and in consequence, $x < \frac{4}{5}$ always. And since $u_1 > \frac{1}{6}$ and $u_2 > \frac{1}{6}$ are dominated in the static game, one may expect $u_1 + u_2 \le \frac{1}{3}$ whenever prevailing prices are low, so that $x > \frac{2}{3}$ always. In short, it really seems necessary to define good strategies $u_1^*(\cdot)$ and $u_2^*(\cdot)$ only for $\frac{2}{3} < x < \frac{4}{5}$. And in that interval, the choices $u_1(x) = u_2(x) = u^*(x)$ drive $x(t)$ to $x^* = \frac{3}{4} = x_\infty$ from any initial position. For values of x outside that interval, it seems adequate to stipulate just that u_i should be large when $x > \frac{4}{5}$ and small when $x < \frac{2}{3}$.

Whatever practical purpose is to be served by the analysis of the game at hand has doubtless been served by now. But we shall pursue it longer because it is one of the few we know that is simple enough to permit a solution for $\rho > 0$.

Substituting $u(x) = y(x) + \frac{1}{2} P_1(x)$ into (4.8) and differentiating once yields

(4.13) $$2y(x)y'(x) = P'(x) - \rho\sigma y(x) = Ax - B - \rho\sigma y(x)$$

where $\sigma = 1/(N - \frac{1}{2})$ and

(4.14) $$P(x) = \rho R_1(x) + \frac{1}{4} P_1^{\,2}(x) - \frac{\rho\sigma}{2} \int_{x_0}^{x} P_1(s)\,ds - Q_1(x) .$$

So if one defines

(4.15) $$\begin{aligned} r_1, r_2 &= (-\rho\sigma \pm \sqrt{\rho^2\sigma^2 + 8A})/2 \\ c_1 &= (r_2(x_0 - B/A) - 2y_0)/(r_2 - r_1) \\ c_2 &= (2y_0 - r_1(x_0 - B/A))/(r_2 - r_1) \end{aligned}$$

one need only eliminate t between the relations

(4.16) $$\begin{aligned} x &= c_1 e^{r_1 t} + c_2 e^{r_2 t} + B/A \\ y &= \tfrac{1}{2} c_1 r_1 e^{r_1 t} + \tfrac{1}{2} c_2 r_2 e^{r_2 t} \end{aligned}$$

to obtain the solution of (4.8) thru (x_0, y_0). For the expressions (4.16) for x and y satisfy

(4.17) $\dot{x} = 2y$ and $\dot{y} = Ax - B - \rho\sigma y$.

The result of such a computation is shown also in figure 4.2 for the parameter values $N = 2$, $a = 1$, $c = \frac{1}{2}$, $\rho = \frac{1}{10}$. It is interesting how little the optimal strategy $u^*(\cdot)$ changes in response to the increase in ρ .

It is of interest too to mention certain computational methods which do not work for the present game. The first rests on the fact that (4.3) is an instance of the general linear-quadratic game, so called because $\dot{x}$ equals a linear form in the $N + 1$ variables $x, u_1, \ldots, u_N$ while the payoff integrands all are quadratic in those variables. And the Hamilton-Jacobi equations of a linear-quadratic game ordinarily have quadratic solutions, which may be found by the method of undetermined coefficients. But if $V_i(x)$ is quadratic in x, then $V_i'(x)$ is linear and so is $u_i(x)$ by (4.5). The strategies corresponding to quadratic solutions of the Hamilton-Jacobi equations in the present game, for instance, may be found by substituting $y = mx - b$ in (4.13). There are exactly two of them. But the method can never yield the highly non-linear strategies $u^*(x)$ depicted in figure 4.2. Generally speaking, the linear equilibrium strategy pairs will be poor ones for the associatal static game (4.12).

Another method of solving the problem at hand which immediately suggests itself is simply to expand the solutions of (4.13) in their Taylor series about $x = x_0$. For by differentiating the equations again and again, one obtains successively

$$\begin{aligned} 2yy'' &= A - \rho\sigma y'^2 , \\ 2yy''' &= -(\rho\sigma y'' + by'y'') , \\ 2yy^{iv} &= -(\rho\sigma y''' + 8y'y''' + by''^2) , \\ 2yy^{v} &= -(\rho\sigma y^{iv} + 10y'y^{iv} + 20\, y''y''') , \end{aligned} \tag{4.18}$$

etc. But $y_0 = u_0 - P_1/2 = -\frac{1}{24}$ in the case $N = 2$, $a = 1$, $c = \frac{1}{2}$ and $\rho = \frac{1}{10}$, so the Taylor-series coefficients for the expansion about $x_0 = \frac{3}{4}$ are $a_0 = y_0 = -.04167$, $a_1 = -.1$, $a_2 = -8.32$, $a_3 = 17.75$, $a_4 = 791.62$, $a_5 = -5317.43$, $a_6 = -142236.11, \ldots$ which do not quickly become small, as they must if the series is to converge rapidly. And it does not. For instance, at $x = .85$ the partial sums are successively .115, .032, .041, .047, .047, .047, and do not converge quickly to the correct value of about .102. In short, little is to be expected of this sort of series expansion when more complex problems are undertaken.

More special series expansions, however, seem more promising. For instance, an expansion of the form

$$(4.19)\qquad y(x) = y_0(x) + \rho y_1(x) + \rho^2 y_2{}^2(x) + \ldots$$

in the parameter ρ seems natural, because ρ is so often small in cases of interest. Trying (4.19) in the integrated form

$$(4.20)\qquad y^2(x) = P_0(x) + \rho Q_0(x) - \rho\sigma \int_{x_0}^{x} y(s)\,ds\ ,$$

of (4.13), where now

$$(4.21)\qquad P(x) = P_0(x) + \rho Q_0(x) = \left(\tfrac{1}{4} P_1{}^2(x) - Q_1(x)\right) + \rho\left(R_1(x) - \sigma\int_{x_0}^{x} P_1(s)\,ds\right)\ ,$$

yields successively

$$(4.22)\qquad y_0(x) = \pm\sqrt{P_0(x)}\ ,\quad y_1(x) = \frac{-1}{2y_0(x)}\left(Q_0(x) + \sigma\int_{x_0}^{x} y_0(s)\,ds\right)\ ,$$

etc. the indeterminate sign being again negative since $y_0 = y(x_0) = y_0(x_0) < 0$. There can be no doubt that the present method works better than the previous one for this game, since $y_0(x)$ is already a better approximation to $y(x)$ than even the sixth partial sum of its Taylor expansion about $x_0 = \frac{3}{4}$. And $y_0(x) + \rho y_1(x)$ is an excellent approximation for rather larger values of ρ than $\frac{1}{10}$. But additional coefficients $y_2(x), y_3(x), \ldots$ can only be determined by numerical integration, as the closed form of $y_1(x)$ is not simple.

The method of expansion in series of the form (4.19) seems by far the best we know for solving the relatively complex and non-linear games which are of greatest interest for economic applications. However the method as presently constituted is very special in that it applies only to problems with a single state variable. An extension of the method to problems of higher dimension would represent a very important advance indeed for the theory.

5. A Linearized Version. In what has gone before, we have largely been concerned with the effects of non-linearity. Such was invariably the case for the static games considered, and it was the case also for the differential game discussed because of the

quadratic terms which appear in the payoff integrands. When the players' action variables appear only <u>linearly</u> in the model, the nature of the games and of their solutions differ markedly. To illustrate, we consider the game (5.3) in the case $N = 2$ and $a_1 = a_2 = 0$.

The Hamiltonian functions now become

$$(5.1)\qquad H_1 = (x - c_1 - P_1)\, u_1 + P_1(1 - x - u_2)$$

$$P_2 = (x - c_2 - P_2)\, u_2 + P_2(1 - x - u_1)\ ,$$

and are maximized by the choices

$$(5.2)\qquad \hat{u}_i(x, P_i) = m \quad \text{if} \quad P_i < x - c_i$$

$$= 0 \quad \text{if} \quad P_i > x - c_i\ ,$$

$$\text{indeterminate} \quad \text{if} \quad P_i = x - c_i\ ,$$

m being some upper bound on the rate of production. There is no reason m must be the same for both players, but we assume it so for simplicity. We also assume m very large, so that it will not later prove restrictive. The cases $P_i \neq x - c_i$ seem uninteresting, as they arise only during periods of zero or peak production. Hence we shall investigate first the cases in which $P_i = x - c_i$.

Recalling that $P_i = V_i'(x)$, it is clear that $P_i = x - c_i$ implies

$$(5.3)\qquad V_i(x) = \frac{x^2}{2} - c_i x + k_i\ ,$$

where k_i is an undetermined constant of integration and $i = 1, 2$. So in case $P_1 = x - c_1$ and $P_2 = x - c_2$, the Hamilton-Jacobi equation becomes

$$(5.4)\qquad H_1(x, \hat{u}_1, \hat{u}_2, P_1) = (x - c_1)(1 - x - \hat{u}_2) = \rho V_1(x)$$

$$H_2(x, \hat{u}_1, \hat{u}_2, P_2) = (x - c_2)(1 - x - \hat{u}_1) = \rho V_2(x)\ ,$$

which may be solved independently and uniquely for

$$u_1^0(x) = \hat{u}_1(x, V_1'(x)) = (1 + \rho c_2/2) - (1 + \rho /2)x ,$$
(5.5)
$$u_2^0(x) = \hat{u}_2(x, V_2'(x)) = (1 + \rho c_1/2) - (1 + \rho/2)x.$$

We note the circle of implications

(5.6) $$p_1 = x - c_1 \Rightarrow u_2 = u_2^0(x) \Rightarrow u_2 \notin \{0, m\} \Rightarrow p_2 = x - c_2$$
$$\Rightarrow u_1 = u_1^0(x) \Rightarrow u_1 \notin \{0, m\} = p_1 = x - c_1 ,$$

whereby the intervals on which $u_1 = u_1^0(x)$ and $u_2 = u_2^0(x)$ must coincide if the equations (5.4) and the specifications (5.2) are to remain valid. Remark too that the strategies $u_1^0(x)$ and $u_2^0(x)$ are well and uniquely defined on the interval $0 \le x \le 1$. And $u_i^0(x)$ is non-negative for $x \le (2 + \rho c_i)/(2 + \rho)$. Moreover, if the two are played against one another, they cause $x(t)$ to decrease without bound when $x(0) < x^0 = (1 + \rho\bar{c})/(1 + \rho)$, where $\bar{c} = (c_1 + c_2)/2$, to remain constant if $x(0) = x^0$, and to increase if $x(0) > x^0$. So $u_i^0(x)$ cannot be played indefinitely. For they drive $x(t)$ out of the interval $0 \le x \le 1$ when $x(0) \neq x^0$, and they keep it at that absurd level when $x(0) = x^0$. It is clear that x^0 _is_ an absurd price level because it is nearly one when ρ is small, whereas only dominated production levels u_1 and u_2 can produce such large prices in the static version of the game. One may therefore conclude that, while robust strategies may agree with $u_i^0(x)$ on certain subintervals $a < x < b$ of $0 < x < 1$, they may not do so for x near 0 or 1.

Accordingly it seems natural to seek equilibrium pairs $(u_1^*(x), u_2^*(x))$ of strategies which agree with $(u_1^0(x), u_2^0(x))$ in the middle portion of $0 \le x \le 1$, while assuming extreme values at the ends of that interval.

In an interval $a < x < b$ wherein u_1 and u_2 are constant, the Hamilton-Jacobi equations take the form

(5.7) $$(x - c_i) u_i + (1 - x - u_1 - u_2) V_i'(x) = \rho V_i(x) ,$$

$i = 1,2$. So differentiating once and setting $p_i(x) = V_i'(x)$, one obtains

(5.8) $$\frac{dp_i(x)}{dx} = \frac{(1+\rho)p_i-u_i}{1-x-u_1-u_2} .$$

The derivatives of the value functions $V_1(x)$ and $V_2(x)$ must be solutions of (5.8), wherein u_i and u_2 are either 0 or m <u>except</u> on subintervals of the unit interval whereon $p_i(x) = V_i'(x) = x - c_i$ for $i = 1, 2$. Hence the situation is as shown in figure 4.3; the graph of $V_i'(x)$ coincides with the line $p_i = x - c_i$ in the middle portion of $0 \leq x \leq 1$,

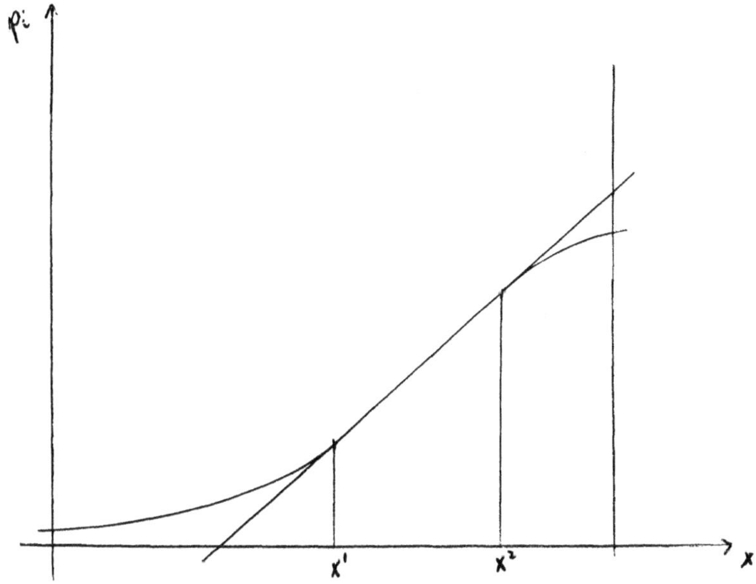

Figure 4.3

with a solution of (5.8) for which $u_1 = u_2 = m$ to the right of that interval, and with another for which $u_1 = u_2 = 0$ to the left of it. The last mentioned solutions of (5.8) do in fact lie below and above the line $p_i = x - c_i$ as shown, provided only that $x^1 \leq (1 + (1 + \rho)c_i)/(2 + \rho)$ and $m > 1/2$.

The curve indicated in figure 4.3, of course, represents an entire continuum of solution pairs $(V_1(x), V_2(x))$ of the Hamilton-Jacobi equations, because constants of integration are introduced in passing from $V_i'(x)$ to $V_i(x)$ <u>and</u> because the "nodes" x^1 and x^2 at which the various local solutions are pieced together into a global one may vary continuously. The additive constants of integration are of little interest here, as the equilibrium strategies $u_i^*(x) = \hat{u}_i(x, V_i'(x))$ are independent of them. But the nodes x^1 and x^2 are of critical import.

To understand the roles of x^1 and x^2, one need only glance at the field of trajectories $x(t)$, $0 \leq t < \infty$ generated by the equilibrium strategies defined thru (5.2) by a pair of solutions $(V_1(x), V_2(x))$ of the Hamilton-Jacobi equations. They are as shown in figure 4.4, and assume the value x^1 after a finite time.

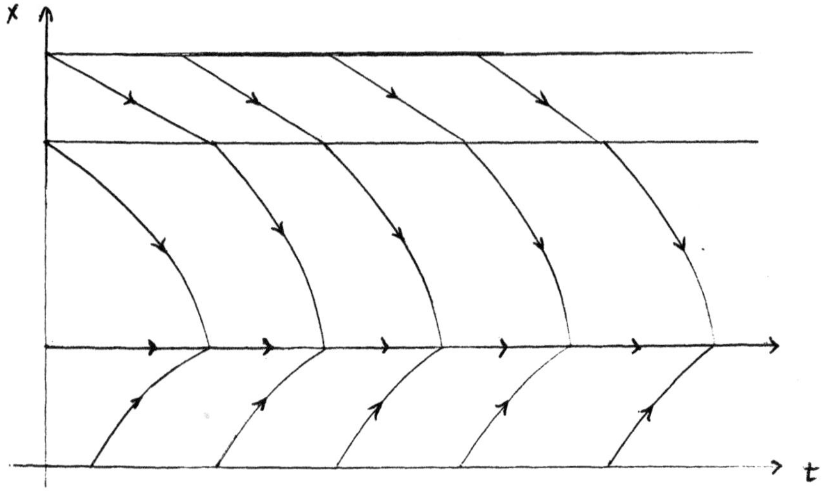

Figure 4.4

So x^1, whatever it is and however determined, is the steady state price which results from the equilibrium strategy pair $u_1^*(x)$, $u_2^*(x)$ generated by a pair of solutions $V_1(x)$, $V_2(x)$ of the Hamilton-Jacobi equations. There can be no doubt, incidentally, that such a pair of strategies is in Nash equilibrium, since the graph of $V_i'(x)$ shown in figure 4.3 is continuous. For then $V_1(x)$ and $V_2(x)$ are continuously differentiable on the entire interval $0 < x < 1$, and each $u_i^*(x)$ is optimal against the other. But they do have certain other shortcomings.

To see this, find an equilibrium triple $(x,\bar{u}_1,\bar{u}_2)$ for the static game

$$(5.9)\qquad 1 \ \underset{0\le u_1\le m}{\text{maximize}}\ (x - c_1)u_1 \qquad 2 \ \underset{0\le u_2\le m}{\text{maximize}}\ (x - c_2)u_2$$
$$\text{subject to } x = 1 - u_1 - u_2$$

The method of Lagrange multipliers yields quickly

$$(5.10)\qquad \begin{aligned} \bar{x} &= (1 + c_1 + c_2)/3 \\ \bar{u}_1 &= (1 - 2c_1 + c_2)/3 \\ \bar{u}_2 &= (1 + c_1 - 2\, c_2)/3 \, . \end{aligned}$$

But

$$(5.11) \qquad u_1^0(\bar{x}) = (1 - 2c_1 + c_2)/3 + (1 - \rho/2)(1 - c_1 + 2c_2)/3,$$

which equals $\bar{u}_1$ only if $\rho = 2$. For other values of ρ, and in particular for the important small values of ρ, the two are quite different. In the case $c_1 = c_2 = 1/2$, $\rho = 1/10$ for instance, $\bar{u}_1(x) = 13/40$ which is almost twice $\bar{u}_1 = 1/6$. So even if $x^1 = \bar{x}$, players employing the equilibrium strategies

$$(5.12) \qquad \begin{aligned} u_i^*(x) &= m & &\text{if} \quad x \geq x^2 \\ &= u_1^0(x) & &\text{if} \quad x^2 > x > x^1 \\ &= 0 & &\text{if} \quad x^1 > x \end{aligned}$$

will not produce at anything like the rates $\bar{u}_1$, $\bar{u}_2$, as $x(t)$ nears its equilibrium position x^1. Rather they will produce at the much greater rates $u_1^0(x)$ for $x > x'$, and will not produce at all when $x < x'$. It is not clear what rates they will produce at when $x = x^1$.

To resolve that question, one observes that the knowledge $u_i^*(x) = u_i^0(x)$ for $x > x^1$ and $u_i^*(x) = 0$ for $x < x^1$ determines $u^*(x^1)$ and $u_2^*(x^1)$ in a natural fashion. For if $x(t)$ be imagined to oscillate minutely about the line $x = \bar{x} = x^1$, instead of lying always on it, the players would produce at roughly the rates $u_1^0(\bar{x})$ and $u_2^0(\bar{x})$ while $x(t) > \bar{x}$, and would not produce while $x(t) < \bar{x}$. So the effect achieved would be identical with that of defining $u_1^*(\bar{x})$ and $u_2^*(\bar{x})$ to be the solutions of the simultaneous equations $u_1/u_2 = u_1^0(\bar{x})/u_2^0(\bar{x})$ and $u_1 + u_2 = 1 = \bar{x}$. Then $u_1^*(\bar{x})$ and $u_2^*(\bar{x})$ are substantially smaller than $u_1^0(\bar{x})$ and $u_2^0(\bar{x})$ since $1 - \bar{x} - u_2^0(\bar{x}) - u_1^0(\bar{x})$ is negative, and consequently each $u_i^*(x)$ is discontinuous as $\bar{x}$ approaches $x = x^1$ from either side.

To appreciate fully the violence that such a conclusion does one's economic intuition, it is instructive to consider a pair of firms introducing a new product into the market. Recent experience (with jet air travel, color T.V., and pocket calculatiors for instance) suggests that initially the supply will be small and the price dear. Then gradually the supply will increase and the price decline until a steady-state is achieved. But that is not at all what would be observed if the game were played according to the strategies (5.12). Instead total production would rise gradually to the level $u_1^0(\bar{x}) + u_2^0(\bar{x})$ as $x(t)$ came near to x, then drop precipitously to the lower level $u_1^*(\bar{x}) + u_2^*(\bar{x}) = 1 - \bar{x}$. And that drop in production would not be followed by any price rise!

Such phenomena are ubiquitous in games wherein the action variables appear only linearly; their solutions are often in terms of "bang-bang" strategies (which take only extreme values) and nearly always in terms of discontinuous strategies of the sort described above. So while such games are perhaps mathematically intriguing, it is not at all clear that their solutions can be expected to convey valuable economic insight. Indeed the opposite would appear to be the case in the present example.

THE EXISTENCE OF OPTIMAL STRATEGIES AND SADDLE POINTS IN STOCHASTIC DIFFERENTIAL GAMES

Robert J. Elliott,
University of Hull, England.

1. INTRODUCTION

In notable papers [2], [3], M. H. A. Davis and P. Varaiya used martingale methods to discuss the optimal control of a stochastic system. The dynamics of the system were described by a stochastic differential equation whose solution was interpreted by the Girsanov measure transformation method. In [4] the present author applied the techniques of Davis and Varaiya to a two person zero sum stochastic differential game. Later, using martingale representation results and the Doob-Meyer decomposition of submartingales, the optimal control results of Davis and Varaiya were simplified and extended in [5] and [6].

The object of the present paper is to apply the more simple methods of [5] and [6] to the situation of a two person zero sum stochastic differential game, and also, by using a more natural Hamiltonian, to clarify the results of [2] and [6] on the existence of optimal controls. Furthermore, by these direct methods we re-establish the results of [4].

As in [4], the evolution of the system is described by a stochastic differential equation and associated with the process is a real valued cost, or payoff. There are now two controllers, or players, J_1 and J_2. As the game is zero-sum we suppose J_1 is choosing his control to maximize the payoff and J_2 is choosing his control to minimize the payoff. The problem investigated below is how J_1 and J_2 should select their controls.

2. DYNAMICS AND PAYOFF

Suppose the dynamics are described by a stochastic differential equation

$$dx_t = f(t,x,y,z)dt + \sigma(t,x)dB_t ,$$

with initial condition $x(o) = x_o \in R^m$. Here $t \in [o,1]$ and B is an m-dimensional Brownian motion. Write C for the space of continuous functions from $[o,1]$ to R^m, and $F_t = \sigma\{x_s : s \leq t\}$ for the σ-field generated on C up to time t. We suppose f, which maps $[0,1] \times C \times Y \times Z$ into R^m, and the nonsingular $m \times m$ matrix σ, defined on $[0,1] \times C$, satisfy the usual measurability and growth conditions (see [3]). Given an m-dimensional Brownian motion B_t on a probability space (Ω, μ), these conditions on σ ensure that the equation

$$x_t = x_0 + \int_0^t \sigma(s,x)dB_s$$

has a unique solution with sample paths in C. Write $A_t = \sigma\{B_s : s \leq t\}$ for the σ-field on Ω generated by B_s up to time t. Then, as a process on Ω, x_t is A_t adapted and, because $dB = \sigma^{-1}dx$ and σ is Lipschitz, we see that $A_t = x^{-1}(F_t)$.

The spaces Y and Z, where the control functions take values, are compact metric spaces, and we suppose f is continuous in the variables $y \in Y$ and $z \in Z$. The admissible feedback controls M_s^t for the player J_1, over $[s,t] \subset [0,1]$, are the measurable functions $y : [s,t] \times C \to Y$ (Y is given the Borel σ-field), such that for each τ, $s \leq \tau \leq t$, $y(\tau, \cdot)$ is F_t measurable and for each $x \in C$ $y(\cdot, x)$ is Lebesgue measurable. The admissible feedback controls N_s^t for J_2 over $[s,t] \subset [0,1]$ are the measurable functions $z : [s,t] \times C \to Z$ with similar properties. For $y \in M_s^t$ and $z \in N_s^t$, write

$$f^{y,z}(\tau, x) = f(\tau, x, y(\tau, x), z(\tau, x)) .$$

Then the conditions on f ensure that

$$E[\exp \xi_s^t (f^{y,z}) | F_s] = 1 \text{ a.s. } \mu$$

where

$$\xi_s^t(f^{y,z}) = \int_s^t \{\sigma^{-1}(\tau, x)f^{y,z}(\tau, x)\}' dB_\tau - 1/2 \int_s^t |\sigma^{-1}(\tau, x)f^{y,z}(\tau, x)|^2 d\tau .$$

Write $M = M_0^1$ and $N = N_0^1$. For each $y \in M$ and $z \in N$ a probability measure $\mu_{y,z}$ is defined on Ω by putting

$$d\mu_{y,z}/d\mu = \exp \xi_0^1 (f^{y,z}) .$$

Girsanov's theorem then states the following:

THEOREM 2.1. Under the measure $\mu_{y,z}$ the process $w_t^{y,z}$ is a Brownian motion on Ω, where

$$dw_t^{y,z} = \sigma^{-1}(t,x)(dx_t - f^{y,z}(t,x)dt) .$$

PAYOFF. The payoff is supposed to be of the form

$$g(x(1)) + \int_0^1 h(t,x,y,z)dt \,,$$

where g and h are real valued, and bounded, $g(x(1))$ is F_1 measurable and h satisfies the same conditions as the components of f. Corresponding to controls $y \in M$ and $z \in N$ the expected total cost is

$$P(y,z) = E_{y,z}[g(x(1)) + \int_0^1 h_t^{y,z} dt]$$

where

$$h_t^{y,z} = h(t,x,y(t,x),\ z(t,x)) \,,$$

and $E_{y,z}$ denotes expectation with respect to $\mu_{y,z}$. We are considering a zero sum differential game, so J_1 wishes to choose y so that $P(y,z)$ is maximized and J_2 wishes to choose z so that $P(y,z)$ is minimized.

3. PRINCIPLE OF OPTIMALITY

Suppose J_2 plays the control $z(t,x) \in M$ throughout the game. Then if J_1 plays the control $y(t,x) \in N$, the cost incurred from time t onwards, given F_t, is independent of the controls used up to time t and is

$$\psi_t^{y,z} = E_{y,z}[g(x(1)) + \int_t^1 h_s^{y,z}\, ds | F_t] \,.$$

Because $L^1(\Omega)$ is a complete lattice, the supremum

$$W_t^z = \bigvee_{y \in M} \psi_t^{y,z}$$

exists, and represents the best that J_1 can attain from t onwards, given that J_2 is using control $z \in N$. Re-phrasing Theorem 3.1 of [3] we have

THEOREM 3.1. (i) $y^*(z)$ is the optimal reply to z if and only if

$$W_t^z + \int_0^t h_s^{y^*(z),z}\, ds$$

is a martingale on $(\Omega, A_t, \mu_{y^*(z),z})$.

(ii) in general, for $y \in M$,

$$W_t^z + \int_0^t h_s^{y,z}\, ds$$

is a supermartingale on $(\Omega, A_t, \mu_{y,z})$.

From the martingale representation results of [1], therefore, we have that $y^*(z)$ is the optimal reply for J_1 to z if and only if there is a predictable process g_t^z such that

$$\int_0^1 |g_s^z|^2 ds < \infty \text{ a.s.}$$

and

$$W_t^z + \int_0^t h_s^{y^*(z),z} ds = J^z + \int_0^t g_s^z dw_s^{y^*(z),z} .$$

Here $J^z = W_0^z$ and the last integral is a stochastic integral with respect to the Brownian motion $w^{y^*(z),z}$ on $(\Omega, \mu_{y^*(z),z})$, as defined in Girsanov's theorem above. For any other $y \in M$ the supermartingale $W_t^z + \int_0^t h^{y,z} ds$ has a unique Doob-Meyer decomposition as $J^z + M_t^{y,z} + A_t^{y,z}$, where $M_t^{y,z}$ is a martingale on $(\Omega, \mu_{y,z})$ and $A_t^{y,z}$ is a predictable decreasing process. From the above representation, though,

$$W_t^z + \int_0^t h_s^{y,z} ds = J^z + \int_0^t g^z \sigma^{-1}(dx_s - f_s^{y,z} ds)$$

$$- \int_0^t (g^z \sigma^{-1} f_s^{y^*(z),z} + h_z^{y^*(z),z}) - (g^z \sigma^{-1} f_s^{y,z} + h_s^{y,z}) ds .$$

Again from Theorem 2.1 $dw_s^{y,z} = \sigma^{-1}(dx_s - f_s^{y,z} ds)$ is a Brownian motion on $(\Omega, \mu_{y,z})$, and so the stochastic integral with respect to $dw_s^{y,z}$ is a martingale. The second integral is a predictable process, so by the uniqueness of the Doob-Meyer decomposition

$$M_t^{y,z} = \int_0^t g^z dw^{y,z} ,$$

$$A_t^{y,z} = \int_0^t (g^z \sigma^{-1} f_s^{y^*(z),z} + h_s^{y^*(z),z}) - (g^z \sigma^{-1} f_s^{y,z} + h_s^{y,z}) ds .$$

Because $A_t^{y,z}$ is decreasing we have the following principle of optimality:

THEOREM 3.2. If $y^*(z)$ is the best reply for J_1 then, almost surely,

$$g^z \sigma^{-1} f_s^{y^*(z),z} + h_s^{y^*(z),z} \geq g^z \sigma^{-1} f_s^{y,z} + h_s^{y,z} .$$

That is, if the optimal reply for J_1 exists, it is obtained by maximizing the

Hamiltonian

$$g^z \sigma^{-1} f_s^{y,z} + h_s^{y,z} .$$

4. THE EXISTENCE OF OPTIMAL CONTROLS

We now establish the existence of an optimal control $y^*(z) \in M$ for J_1 in reply to any control $z \in N$ played by J_2.

As in the above section, we suppose that it is known that J_2 uses control $z \in N$ throughout the game. Then the process W_t^z exists whether or not there is an optimal control $y^*(z)$. Improving and extending the results of [6] we first obtain an integral representation for this process. To do this, as in [6], it is convenient to make the payoff into a completely terminal payoff by introducing a new state variable x_{m+1} and a new Brownian motion B_{m+1} on a probability space (Ω', μ'). Suppose x_{m+1} satisfies the stochastic equation

$$dx_{m+1} = h(t,x,y,z)dt + dB_{m+1}$$

$$x_{m+1}(o) = o .$$

The $(m+1)$ dimensional process (x, x_{m+1}) is defined on the product space

$$(\Omega^+, \mu^+) = (\Omega \times \Omega', \mu \times \mu') .$$

In fact, writing $x^+ = (x, x_{m+1})$ we have $dx^+ = f^+ dt + \sigma^+ dw^+$ where $f^+ = (f,h)$, $\sigma^+ = \begin{bmatrix} \sigma & o \\ o & 1 \end{bmatrix}$ and, with $w_{m+1} = B_{m+1}$, $w^+ = (w, w_{m+1})$ is an $m+1$ dimensional Brownian motion on Ω^+.

Corresponding to controls $y \in M$, $z \in N$ a new probability measure $\mu_{y,z}^+$ is defined on Ω^+ by putting

$$d\mu_{y,z}^+ / d\mu^+ = \exp \xi_o^1 (f_{y,z}^+)$$

where

$$f_{y,z}^+ = (f^{y,z}, h^{y,z}) .$$

If $E_{y,z}^+$ denotes the expectation with respect to $\mu_{y,z}^+$ the expected payoff corresponding to controls y and z is

$$E^+_{y,z}[g(x(1)) + x_{m+1}(1)]$$

$$= E_{y,z}[g(x(1)) + \int_0^1 h_s^{y,z} ds] ,$$

because w_{m+1} is a Brownian motion and h and g are independent of x_{m+1}.

As in [5] define

$$W^+_z(t) = \bigvee_{y \in M} E^+_{y,z}[g(x(1)) + x_{m+1}(1) | F^+_t] ,$$

the supremum now being in $L^1(\Omega^+)$. Here we are supposing that C^+ denotes the R^{m+1} valued continuous functions on $[o,1]$ and F^+_t the σ-field on C^+ generated up to time t. Write Φ^+ for the set of functions

$$\phi : [o,1] \times C^+ \to R^{m+1}$$

which satisfy

(i) for each $t \in [o,1]$ $\phi(t,\cdot)$ is F^+_t measurable,

(ii) for each $x \in C^+$ $\phi(\cdot,x)$ is Lebesgue measurable,

(iii) $|(\sigma^+)^{-1}(t,x)\phi(t,x)| \leq M(1 + \|x\|_t)$ where $\|x\|_t = \sup_{o \leq s \leq t} |x(s)|$.

Write, as in [2], $\mathcal{D} = \{\exp \xi^1_o(\phi) : \phi \in \Phi^+\}$ so that because ϕ has linear growth $E^+ \exp \xi^1_o(\phi) = 1$ for all $\phi \in \Phi^+$, where E^+ denotes expectation with respect to μ^+. Note that for any $y \in M$, $z \in N$

$$f^+_{y,z} = (f^{y,z}, h^{y,z}) \in \Phi^+ .$$

Then from Theorem 9.3 of [6], because $\mathcal{D}$ is weakly compact, we have the following result:

THEOREM 4.1. There is a function $H \in \Phi^+$ such that $(W^+_z(t), F^+_t, \mu^*)$ is a martingale. Here μ^* is defined on Ω^+ by

$$\frac{d\mu^*}{d\mu^+} = \exp \xi^1_o(H) .$$

Of course, if there is an optimal reply $y^*(z)$ for J_1, as shown in section 2 we can take $H = f^+_{y^*(z),z}$. The above result states that, even if there not an optimal control, there is always a 'drift term' $H \in \Phi^+$ whose corresponding measure gives the maximum value function

$$W_z^+(t) = \bigvee_{y \in M} E_{y,z}^+[g(x(1)) + x_{m+1}(1) | F_t^+]$$

as

$$= E^*[g(x(1)) + x_{m+1}(1) | F_t^+] \ .$$

(E^* denotes expectation with respect to μ^*).

Under μ^*, using Girsanov's theorem, we are considering an $m + 1$ dimensional Brownian motion w^* on (Ω^+, μ^*) defined by

$$\begin{pmatrix} dw^* \\ dw^*_{m+1} \end{pmatrix} = \begin{pmatrix} \sigma^{-1} & 0 \\ 0 & 1 \end{pmatrix} \begin{pmatrix} dx & - \hat{H}\, dt \\ dx_{m+1} & - H_{m+1} dt \end{pmatrix} .$$

Here $\hat{H}$ denotes the first m coordinates of H .

Because $h(t,x,y(t,x), z(t,x))$ is independent of x_{m+1} , for any controls, the weak limit H_{m+1} is independent of x_{m+1} . Certainly $g(x(1))$ is independent of x_{m+1} , so for any control $y \in M$:

$$E_{y,z}^+[g(x(1)) + \int_0^1 h(s,x,y,z)ds + w_{m+1}(1) - w_{m+1}(t) | F_t^+] + x_{m+1}(t)$$

$$= E_{y,z}[g(x(1)) + \int_t^1 h(s,x,y,z)ds | F_t] + x_{m+1}(t).$$

Taking the supremum to obtain $W_z^+(t)$ we see

$$W_z^+(t) = W_t^z + \int_0^t H_{m+1}(s)ds + w^*_{m+1}(t) \ ,$$

where W^z is the F_t adapted process defined in section 3. Therefore,

$$W_t^z + \int_0^t H_{m+1}(s)ds + w^*_{m+1}(t) = E^*[g(x(1)) + x_{m+1}(1) | F_t^+] \ .$$

Taking expectations with respect to $F_t \subset F_t^+$ we have

$$W_t^z + \int_0^t H_{m+1}(s)ds = E^*[g(x(1)) + x_{m+1}(1) | F_t]$$

Therefore, $W_t^z + \int_0^t H_{m+1}(s)ds$ is a martingale on (Ω, A_t, μ^*) , and so can be represented as a stochastic integral, $J^z + \int_0^t g^* dw^*$, with respect to the m-dimensional Brownian motion w^* defined on (Ω, A_t, μ^*) by

$$dw^* = \sigma^{-1}dx - \sigma^{-1}H\, dt \quad .$$

Here $J^z = W_0^z$ and g^* is a predictable adapted process. Under any other control $y \in M$, as in Theorem 3.1, $W_t^z + \int_0^t h_s^{y,z} ds$ is a supermartingale and so arguing as before

$$W_t^z + \int_0^t h_s^{y,z}\, ds = J^z + \int_0^t g^*\, dw_s^{y,z} + \int_0^t (g^*\sigma^{-1} f_s^{y,z} + h_s^{y,z}) - (g^*\sigma^{-1}\hat{H}_s + H_{m+1}(s))ds .$$

As in section 3, $w_s^{y,z}$ is the Brownian motion on $(\Omega, \mu_{y,z})$ defined by

$$d\,w_s^{y,z} = \sigma^{-1}(dx_s - f_s^{y,z} ds) .$$

The first integral on the right hand side is again a stochastic integral and the second a decreasing process, so we have that almost surely

$$g^*\sigma^{-1}\hat{H} + H_{m+1} \geq g^*\sigma^{-1} f^{y,z} + h^{y,z} \tag{4.1}$$

Note that if there is a process $y^*(z)$ such that, almost surely,

$$g^*\sigma^{-1}\hat{H} + H_{m+1} = g^*\sigma^{-1} f^{y^*(z),z} + h^{y^*(z),z}$$

then

$$W_t^z + \int_0^t h_s^{y^*(z),z} ds = J^z + \int_0^t g^* dw^+_{y^*(z),z}$$

and so is a martingale. Therefore, from Theorem 3.1, $y^*(z)$ would be an optimal reply to z.

For the above process g^*, because f and h are continuous in the control variables y and z and the control spaces are compact, there is as in [4] a measurable feedback control $y^*(z)$ such that almost surely

$$g_s^*\sigma^{-1} f^{y^*(z),z} + h^{y^*(z),z} \geq g_s^*\sigma^{-1} f^{y,z} + h^{y,z} .$$

Using the methods of [2] we now show $y^*(z)$ is an optimal reply for J_1.

Write $\Gamma_s(y,z) = g_s^*\sigma^{-1} f_s^{y,z} + h_s^{y,z}$

$$\hat{\Gamma}_s = g_s^*\sigma^{-1}\hat{H}_s + H_{m+1}(s) ,$$

and suppose $y^*(z)$ is selected as above so that $\Gamma_s(y^*(z),z) \geq \Gamma_s(y,z)$.

Then

$$W_t^z + \int_0^t h_s^{y,z} ds = J^z + \int_0^t g^* dw^+_{y,z} + \int_0^t (\Gamma_s(y,z) - \hat{\Gamma}_s ds .$$

Taking expectations with respect to $\mu^+_{y,z}$ at $t = 1$:

$$E^+_{y,z}[g(x(1)) + \int_0^1 h_s^{y,z} ds] = J^z + E^+_{y,z}[\int_0^1 (\Gamma_s(y,z) - \hat{\Gamma}_s)ds]$$

$$\leq J^z + E^+_{y,z}[\int_0^1 (\Gamma_s(y^*(z),z) - \hat{\Gamma}_s)ds] \quad .$$

The left hand side of the inequality is just $\psi_o^{y,z}$, so for any $n \in Z^+$ there is a control $y_n \in M$ such that

$$- E^+_{y_n,z}[\int_o^1 (\Gamma_s(y^*(z),z) - \hat{\Gamma}_s)ds] < 1/n \; .$$

Writing

$$-X = \int_o^1 (\Gamma_s(y^*(z),z) - \hat{\Gamma}_s)ds$$

the inequality (4.1) implies X is positive almost surely, and $E^+ \phi_n X \to o$, where $\phi_n = \exp \xi_o^1 (f^+_{y_n,z})$. Write $X^N = \min (N,X)$ for $N \in Z^+$,

$$\text{so} \quad o \leq X^N \leq X$$

$$\text{and} \quad E^+ \phi_n X^N \to o \quad .$$

By the weak compactness of $\mathcal{D}$ there is a $\phi \in \mathcal{D}$ such that the ϕ_n converge to ϕ weakly, so

$$\lim_{n\to\infty} E^+ \phi_n X^N = E^+ \phi X^N = o$$

Now $\phi > o$ a.s. so $X^N = o$ a.s. Therefore $X = o$ a.s., and finally

$$\Gamma_s(y^*(z),z) = \hat{\Gamma}_s \text{ a.s.}$$

We can, therefore, conclude that an optimal reply $y^*(z)$ exists for J_1 in reply to any control $z \in N$ played by J_2.

5. THE UPPER VALUE

Similarly to the results of sections 3 and 4 we now establish the existence, and obtain a characterization, of the optimum feedback control that J_2 should use if he chooses his control first. Consequently, we suppose that J_1 will always play his best reply $y^*(z) \in M$ in response to any control $z \in N$ chosen by J_2. The problem we discuss is how J_2, who is trying to minimize the payoff, should choose a $z^* \in N$ such that

$$\inf_{z\in N} \sup_{y\in M} P(y,z) = \inf_{z\in N} P(y^*(z),z) \; .$$

For any $z \in N$ and $t \in [o,1]$, if J_1 plays $y^*(z)$, the expected remaining payoff is

$$\psi_z(t) = E_{y^*(z),z}[g(x(1)) + \int_0^1 h^{y^*(z),z}ds \,|\, F_t] .$$

Because $L^1(\Omega)$ is a complete lattice the infimum

$$V_t^+ = \bigwedge_{z\in N} \psi_z(t)$$

exists in $L^1(\Omega)$.

DEFINITION 5.1. V_t^+ is the upper value function of the differential game, and $V_0^+ = \inf_{z\in N} \sup_{y\in M} P(y,z)$ is the upper value of the game. Again the results of [3] Theorem 3.1 can be adapted to obtain the following:

THEOREM 5.2. (i) $z^* \in N$ is optimal for J_2 if and only if

$$V_t^+ + \int_0^t h^{y^*(z^*),z^*}ds$$

is a martingale on $(\Omega, A_t, \mu_{y^*(z^*),z^*})$.

(ii) in general, for $z \in N$,

$$V_t^+ + \int_0^t h^{y^*(z),z}ds$$

is a submartingale on $(\Omega, A_t, \mu_{y^*(z),z})$.

From the martingale representation results we have, as above, that $z^* \in N$ is optimal for J_2 playing first if and only if there is a predictable process g_t^* such that

$$\int_0^1 |g_s^*|^2 ds < \infty \text{ a.s.}$$

and

$$V_t^+ + \int_0^t h^{y^*(z^*),z^*}ds = J^* + \int_0^t g^* dw_s^* .$$

Here w^* is the Brownian motion given by $dw^* = \sigma^{-1}(dx - f^{y^*(z^*),z^*}ds)$ on $(\Omega, \mu_{y^*(z^*),z^*})$. For general $z \in N$ the submartingale

$$V_t^+ + \int_0^t h_s^{y^*(z),z}ds$$

has a unique Doob-Meyer decomposition $J^* + M_t^z + A_t^z$, where M_t^z is a martingale on $(\Omega, \mu_{y^*(z),z})$ and A_t^z is a predictable increasing process. Similar calculations to those above show that if $z^* \in N$ is optimal for J_2 playing first, then almost surely

$$g_{\cdot}^{*}\sigma^{-1}f_s^{y^*(z^*),z^*} + h_s^{y^*(z^*),z^*} \le g_{\cdot}^{*}\sigma^{-1}f_s^{y^*(z),z} + h_s^{y^*(z),z} \quad .$$

Converseley, arguing as in section 4, without a priori assuming there is an optimal control $z^* \in N$, we can obtain an integral representation for V_t^+, and show that the measurable strategy,obtained by minimizing a Hamiltonian

$$g_{\cdot}^{*}\sigma^{-1}f_s^{y^*(z),z} + h_s^{y^*(z),z} \quad ,$$

exists and is optimal.

Finally we conclude by stating the following result:

THEOREM 5.3. There is a predictable process g^* and $z^* \in N$ is optimal if and only if z^* minimizes the Hamiltonian

$$\Gamma_s(y^*(z),z) = g_{\cdot}^{*}\sigma^{-1}f_s^{y^*(z),z} + h_s^{y^*(z),z} \quad ,$$

(almost surely in (s,ω)).

6. THE ISAACS CONDITION

In definition 5.1 we defined the upper value

$$V_o^+ = \inf_{z\in N} \sup_{y\in M} P(y,z)$$

of the differential game. V_o^+ represents the best outcome that the players J_1 and J_2 can ensure if J_2 chooses his feedback control first. Analogously, we can define the lower value of the game

$$V_o^- = \sup_{y\in M} \inf_{z\in N} P(y,z) \; .$$

We now show that if the Hamiltonian of the game satisfies a saddle-point condition, that is if the upper and lower values of an 'infinitesimal' game are equal, then $V_o^+ = V_o^-$.

DEFINITION 6.1. For $t \in [o,1]$, $x \in C$, $y \in Y$, $z \in Z$ and $p \in R^m$ write

$$L(t,x,p\,;\,y,z) = p.\sigma^{-1}(t,x)f(t,x,y,z) + h(t,x,y,z) \; .$$

The game is said to satisfy the *Isaacs condition* if, for all such t,x,p,

$$\min_{z\in Z} \max_{y\in Y} L(t,x,p\,;\,y,z) = \max_{y\in Y} \min_{z\in Z} L(t,x,p\,;\,y,z) \; .$$

THEOREM 6.2. If the game satisfies the Isaacs condition then $V_0^+ = V_0^-$.

PROOF. Note that for $y \in M$, $z \in N$

$$\Gamma_s(y,z) = L(s,x,g^*; y(t,x), z(t,x))$$

where g^* is the predictable process introduced in section 5.

Furthermore, for any $z \in N$ we showed in section 4 that there exists a strategy $y^*(z) \in M$ such that

$$\Gamma_s(y^*(z),z) = \max_{y\in Y} \Gamma(y,z) \text{a.s.}$$

and then that there is a $z^* \in N$ such that

$$\Gamma_s(y^*(z^*),z^*) = \min_{z\in Z} \Gamma_s(y^*(z),z) \text{a.s.} = \min_{z\in Z}\max_{y\in Y} \Gamma_s(y,z) \quad \text{a.s.}$$

We then had a representation

$$V_t^+ + \int_0^t h^{y^*(z^*),z^*} ds = J^* + \int_0^t g^* dw_s^* \text{ a.s.}$$

However, because f and y are continuous in y and z and Y and Z are compact, as in [4], for any $y \in M$ there exists a strategy $z^*(y) \in N$ such that

$$\Gamma_s(y,z^*(y)) = \min_{z\in Z} \Gamma_s(y,z) \text{ a.s.}$$

Similarly there is a $y^* \in N$ such that

$$\Gamma_s(y^*,z^*(y^*)) = \max_{y\in Y} \Gamma_s(y,z^*(y)) \text{a.s.} = \max_{y\in Y}\min_{z\in Z} \Gamma_s(y,z) \quad \text{a.s.}$$

Because the Isaacs condition holds

$$\Gamma_s(y^*,z^*(y^*)) = \Gamma_s(y^*(z^*),z^*) \text{ a.s.}$$

Now for any $z \in N$

$$\Gamma_s(y^*,z^*(y^*)) \le \Gamma_s(y^*,z) \text{ a.s.}$$

and for any $y \in M$

$$\Gamma_s(y,z^*) \le \Gamma_s(y^*(z^*),z^*) \text{ a.s.}$$

so

$$\Gamma_s(y,z^*) \le \Gamma_s(y^*,z^*) \le \Gamma_s(y^*,z) \text{ a.s.}$$

Therefore

$$V_t^+ + \int_0^t h_s^{y^*,z^*} ds = J^* + \int_0^t g^* dw_s^{y^*,z^*} \quad \text{a.s.}$$

where

$$dw_s^{y^*,z^*} = \sigma^{-1} \, (dx_s - f_s^{y^*,z^*} ds)$$

is a Brownian motion under μ_{y^*,z^*} . For any other $y \in M$:

$$V_t^+ + \int_0^t h_s^{y,z^*} ds = J^* + \int_0^t g^* dw_s^{y,z^*} + \int_0^t (\Gamma_s(y,z^*) - \Gamma_s(y^*,z^*))ds \; .$$

Taking expectations at $t = 1$ with respect to μ_{y,z^*}

$$E_{y,z^*}[g(x(1)) + \int_0^1 h_s^{y,z^*} ds] = P(y,z^*) \le J^* = P(y^*,z^*) \; .$$

Similarly we can show that

$$P(y^*,z^*) \le P(y^*,z) \; .$$

Therefore, if the Isaacs condition is satisfied

$$\sup_{y \in M} \inf_{z \in N} P(y,z) = \inf_{z \in N} \sup_{y \in M} P(y,z) = P(y^*,z^*) = J^* \quad ,$$

that is, the upper and lower values of the differential game are equal. As in [4] we can show that if the upper and lower values are equal then

$$\max_{y \in Y} \min_{z \in Z} L(t,x,g^* ; y,z) = \min_{z \in Z} \max_{y \in Y} L(t,x,g^* ; y,z) \text{ a.s.}$$

REFERENCES

1. CLARK, J.M.C., The representation of functionals of Brownian motion by stochastic integrals, Ann. Math. Stat., 41 (1970), 1282 - 1295.

2. DAVIS, M.H.A., On the existence of optimal policies in stochastic control, S.I.A.M. Jour. Control, 11(1973), 587 - 594.

3. DAVIS, M.H.A. and VARAIYA, P.P., Dynamic programming conditions for partially observable stochastic systems, S.I.A.M. Jour. Control, 11(1973), 226 - 261.

4. ELLIOTT, R.J., The existence of value in stochastic differential games, S.I.A.M. Jour. Control, 14(1976), 85-94.

5. ELLIOTT, R.J., A stochastic minimum principle, Bull. Amer. Math. Soc., 82(1976) 944-946.

6. ELLIOTT, R.J., The optimal control of a stochastic system, S.I.A.M. Jour. Control and Opt.

FEEDBACK STRATEGIES IN DETERMINISTIC DIFFERENTIAL GAMES

Robert J. Elliott,
University of Hull, England.

1. INTRODUCTION

At the conference on Differential Games held at the University of Warwick in 1974, concern was expressed because it appeared difficult to interpret in practice some definitions of strategy being used in differential games. Intuitively, a strategy should be a rule telling a controller, or player, what to do next on the basis of what has happened so far, and in the works of Roxin [6], Varaiya and Lin [7], Friedman [5] and Elliott and Kalton [2], [3], [4], a strategy for a player in a two person zero sum differential game is a non-anticipative function from the control functions used by his opponent into his space of control functions. On the other hand, the feedback strategies discussed in [1] are functions from the trajectory space to the control space, which at any time t are measurable with respect to the information σ-field F_t. The purpose of this short note is to partially resolve the above difficulty in the deterministic case by introducing strategies which are non-anticipative, in fact, "delayed", functions of the trajectory, that is, the strategies are more like feedback ones.

Using the notation defined below, if the dynamics of the system are described by a family of ordinary differential equations

$$\frac{dx}{dt} = f(t,x,y,z),\ x \in R^m,\ t \in [t_o, \infty)\ ,\ x(t_o) = x_o\ ,$$

the usual conditions imposed on the right hand side of such a system, to ensure the existence of a unique solution, are that it is measurable in time t and Lipschitz continuous in x. These conditions, therefore, impose severe restrictions on the kind of feedback controls $y(t,x)$, $z(t,x)$ that can be considered, because they require such controls to be Lipschitz continuous in the state x. This problem is circumvented below by introducing the set of s-delay "feedback" strategies. Two results from [2] are then re-interpreted in terms of these feedback strategies.

2. DIFFERENTIAL GAMES

We first define the dynamics and payoff of the differential game. Suppose Y and Z are compact metric spaces. Consider the dynamical system:

$$\frac{dx}{dt} = f(t,x,y,z) ,$$

where $x \in R^m$ and t, time, belongs to $[t_o,\infty]$. $f : R \times R^m \times Y \times Z \to R^m$ is continuous and satisfies a Lipschitz condition in x :

$$\|f(t,x,y,z) - f(t,x',y,z)\| \leq K\|x - x'\|$$

for $x, x' \in R^m$. The trajectory starts at the initial position x_o at time t_o. A closed subset $F \subset R \times R^m$, such that $F \supset [T,\infty) \times R^m$ for some fixed T, is given and the game ends the first time t_F that the trajectory enters F. The payoff is then of the form

$$P(y,z) = \mu(x(t),\ t_o \leq t \leq t_F) + \int_{t_o}^{t_F} h(t,x(t),y(t),z(t))dt ,$$

where μ is a (non-linear) continuous real valued functional on the Fréchet space

$$C = C([t_o,\infty))^m$$

of continuous functions from $[t_o,\infty)$ to R^m, and $h : R \times R^m \times Y \times Z \to R$ is continuous. We suppose there is a player J_1, controlling $y \in Y$, whose object is to maximize P, and a player J_2, controlling $z \in Z$, whose object is to minimize P.

A dynamical system and payoff as above is called a two person zero sum differential game. This formulation includes the three main types of differential game:-

(i) Survival games. These are discussed in [4], taking $\mu(x) = g(t_F, x(t_F))$ where $g : R \times R^m \to R$ is continuous.

(ii) Pursuit-evasion games. These are discussed in [3]. For this case, if μ is taken to be zero and h is non-negative, so J_2 is always trying to terminate the game, the system is called a generalized pursuit-evasion game. If, further, $h \equiv 1$ the payoff is just the time to capture $t_F - t_o$, so J_1 is the evader and J_2 the pursuer.

(iii) Fixed time games. Taking $F = [T,\infty) \times R^m$ the game always ends at the fixed time T. These games are discussed in the Memoir [2].

3. FEEDBACK STRATEGIES

A map $y : [t_o,\infty) \to Y$ is said to be measurable if for every continuous real valued function ψ on Y, $\psi o y$ is Lebesgue measurable. As above, C denotes the space of continuous functions from $[t_o,\infty)$ to R^m. Write $M_1(t_o)$ (resp. $M_2(t_o)$) for the space of measurable functions from $[t_o,\infty)$ to Y (resp. Z), identifying functions which are equal almost everywhere.

DEFINITION 3.1. An s-delay feedback strategy, $s > 0$ for J_1 is a map

$$\alpha : C \to M_1(t_0)$$

such that whenever

$$x_1(t) = x_2(t), \; t_0 \le t \le t_1$$

then

$$(\alpha x_1)(t) = (\alpha x_2)(t) \text{ a.e.}, \; t_0 \le t \le t_1 + s \; .$$

The set of s-delay feedback strategies for J_1 is denoted by $\Gamma_{t_0}(s)$ ($= \Gamma(s)$ when t_0 is fixed). The s-delay feedback strategies for J_2 are similarly defined and denoted by $\Delta_{t_0}(s) = \Delta(s)$. Note that the values of $(\alpha x)(t)$ for $t_0 \le t \le t_0 + s$ are independent of the trajectory $x(t)$, i.e. for $t_0 \le t \le t_0 + s$ $(\alpha x)(t)$ is a fixed control function $y_\alpha(t) \in M_1(t_0)$.

If $s_1 \ge s_2 > 0$ we have

$$\Gamma(s_2) \supset \Gamma(s_1)$$

$$\Delta(s_2) \supset \Delta(s_1) \; .$$

Each $\alpha \in \Gamma(s)$, $s > 0$, has a value

$$u(\alpha) = \inf \, [P(\alpha x, z) \; ; \; z \in M_2(t_0)] \; ,$$

representing the worst outcome for J_1 if he uses α. Similarly, for $\beta \in \Delta(s)$

$$v(\beta) = \sup \, [P(y, \beta x) \; ; \; y \in M_1(t_0)]$$

DEFINITION 3.2. The upper and lower s-delay values are, respectively,

$$V_s^+(t_0, x_0) = V_s^+ = \inf \, [v(\beta) \; ; \; \beta \in \Delta(s)] \; ,$$

$$V_s^-(t_0, x_0) = V_s^- = \sup \, [u(\alpha) \; ; \; \alpha \in \Gamma(s)] \; .$$

As $s > 0$ gets smaller V_s^+ is monotonic decreasing and V_s^- is monotonic increasing. The limits

$$V^+ = \lim_{s \to 0} V_s^+$$

$$V^- = \lim_{s \to 0} V_s^-$$

therefore exist and define upper and lower values of the game.

LEMMA 3.3. $V^- \leq V^+$.

PROOF. If $\alpha \in \Gamma(s_1)$ and $\beta \in \Delta(s_2)$ then initially α selects a control function $y(t) \in M_1(t_o)$ for $t_o \leq t \leq t_o + s_1$ and β selects a control function $z(t) \in M_2(t_o)$ for $t_o \leq t \leq t_o + s_2$. Therefore, a trajectory $x(t)$ is determined for $t_o \leq t \leq t_o + \min(s_1, s_2)$. The strategies α and β will then determine the control functions over the next time interval, and so on. So the strategies $\alpha \in \Gamma_{t_o}(s_1)$ and $\beta \in \Delta_{t_o}(s_2)$ determine a trajectory $x(t)$ and control functions $y(t)$, $z(t)$, for $t_o \leq t \leq t_F$ and

$$\begin{aligned} P(\alpha,\beta) &= P(\alpha x, z) \\ &= P(y, \beta x) . \end{aligned}$$

Therefore, $u(\alpha) \leq v(\beta)$. So for any $s_1 > 0$, $s_2 > 0$

$$V^-_{s_1} = V^-_{s_1}(t_o, x_o) \leq V^+_{s_2}(t_o, x_o) = V^+_{s_2}$$

so certainly

$$V^- \leq V^+ .$$

4. COMPARISON WITH OTHER VALUES

In previous papers [2], [3], [4], an s-delay strategy for player J_1 is a map

$$\alpha : M_2(t_o) \to M_1(t_o)$$

such that, if $z_1(t) = z_2(t)$ a.e. $t_o \leq t \leq t_1$, then $(\alpha z_1)(t) = (\alpha z_2)(t)$ a.e. $t_o \leq t \leq t_1 + s$. Write $\tilde{\Gamma}(s)$ for the set of such strategies. For $\alpha \in \tilde{\Gamma}(s)$ the values

$$\tilde{u}(\alpha) = \inf\,[P(\alpha z, z)\ ;\ z \in M_2(t_o)]$$

and

$$\tilde{V}^-_s = \sup\,[u(\alpha)\ ;\ \alpha \in \tilde{\Gamma}(s)]$$

are defined.

In the above references an s-delay strategy for J_2 is a map

$$\beta : M_1(t_o) \to M_2(t_o)$$

satisfying a similar condition. $\tilde{\Delta}(s)$ denotes the set of such delay strategies. For $\beta \in \tilde{\Delta}(s)$ write

$$\tilde{v}(\beta) = \sup\,[P(y, \beta y)\ ;\ y \in M_1(t_o)]$$

and

$$\tilde{V}_s^+ = \inf\ [v(\beta)\ ;\ \beta \in \tilde{\Delta}(s)]\ .$$

For any $\alpha \in \Gamma(s)$ and $z \in M_2(t_o)$, α and z determine a trajectory $x(t) \in C$, so $\Gamma(s)$ can be mapped into $\tilde{\Gamma}(s)$. Therefore

$$V_s^- \leq \tilde{V}_s^- \ .$$

However, these two values are, in fact, equal. For any $\varepsilon > o$ suppose $\alpha \in \tilde{\Gamma}(s)$ is such that

$$P(\alpha z, z) \geq \tilde{V}_s^- - \varepsilon$$

for all $z \in M_2(t_o)$. Then we construct $\alpha^* \in \Gamma(s)$ as follows:

Set $\alpha^*(t) = \alpha(t) \in M_1(t_o)$ for $t_o \leq t \leq t_o + s$.

Note we need only define α^* on the subset C^* of C made up of possible solution trajectories of the dynamical system. Suppose $(\alpha^* x)(t)$ is defined for $x \in C^*$ up to time $t_1 \geq t_o + s$. Then if $x \in C^*$ and $z \in M_2(t_o)$ is any control such that $(\alpha^* x, z)$ gives rise to the trajectory $x(t)$ for $t_o \leq t \leq t_1$, define

$$(\alpha^* x)(t) = (\alpha z)(t) \quad \text{for} \quad t_1 < t \leq t_1 + s\ .$$

Using the axiom of choice, we see α^* can be defined for $x \in C^*$ up to time $t_1 + s$. Proceeding in this manner, therefore, there is an s-delay strategy $\alpha^* \in \Gamma(s)$ such that, for any $z \in M_2(t_o)$, (α^*, z) gives the same trajectory as (α, z). Then

$$P(\alpha^* x, z) = P(\alpha z, z) \quad \text{for all} \quad z \in M_2(t_o)\ ,$$

so

$$V_s^- \geq \tilde{V}_s^- - \varepsilon$$

and we can conclude $V_s^- = \tilde{V}_s^-$. Similarly $V_s^+ = \tilde{V}_s^+$, and letting $s \to o$ we have the following result:

THEOREM 4.1. $V^- = \tilde{V}^-$ and $V^+ = \tilde{V}^+$.

For fixed time games we can rephrase Theorem 10.1 of [2] for the values defined by feedback strategies.

THEOREM 4.2. Suppose the Isaacs condition is satisfied, that is for all

$$t,x,p\ :\ \min_z \max_y (p.f(t,x,y,z) + h(t,x,y,z)) = \max_y \min_z (p.f(t,x,y,z) + h(t,x,y,z))\ ,$$

then $V^+ = V^-$.

5. SADDLE POINTS

A sequence $A = (\alpha_n)$ of delay feedback strategies for J_1 will be called an approximate feedback strategy, (i.e. here $\alpha_n \in \bigcup_s \Gamma(\sigma)$ for all n). Similarly an approximate feedback strategy for J_2 is a sequence $B = (\beta_n)$ of delay feedback strategies. We have seen above that for any pair α_n, β_n there is a trajectory $x_n(t)$ and a payoff $P(\alpha_n, \beta_n)$. Denote by $P(A,B)$ the set of all limit points of the sequence $P(\alpha_n, \beta_n)$. Then we say $A^* = (\alpha_n^*)$, $B^* = (\beta_n^*)$ is a saddle point for approximate feedback strategies if for any other approximate feedback strategies $A = (\alpha_n)$, $B = (\beta_n)$

$$P(A,B^*) \leq P(A^*,B^*) \leq P(A^*,B) .$$

Rephrasing Theorem 11.4 of [2] in terms of feedback strategies we have the following result.

THEOREM 5.1. Consider a fixed time differential game, with dynamics and payoff as described in section 2. If the Isaacs condition is satisfied there is a saddle point in approximate feedback strategies.

PROOF. By Theorem 4.2 $V^+ = V^- = V$ so for each $n \in Z^+$ there is a strategy $\alpha_n^* \in \Gamma(1/n)$ and a $\beta_n^* \in \Delta(1/n)$ such that

$$u(\alpha_n^*) = V - \varepsilon_n$$

$$v(\beta_n^*) = V + \delta_n$$

where $\varepsilon_n \to 0$ and $\delta_n \to 0$.

As α_n^* and β_n^* are $1/n$-delay feedback strategies there are controls y_n^* and z_n^*, and a trajectory x_n^*, such that

$$u(\alpha_n^*) \leq P(y_n^*, z_n^*) \leq v(\beta_n^*) .$$

Therefore, $\lim_{n\to\infty} P(y_n^*, z_n^*) = V$. That is, writing $A^* = (\alpha_n^*)$, $B^* = (\beta_n^*)$ we have

$$P(A^*,B^*) = \{V\} .$$

Now, if $A = \{\alpha_n\}$ is any other approximate feedback strategy for J_1 there are controls y_n, z_n, and a trajectory x_n, such that

$$\alpha_n x_n = y_n \text{ and } \beta^* x_n = z_n \text{ for all } n .$$

Therefore, $P(y_n, z_n) \leq v(\beta_n^*) = V + \delta_n$ so $P(A,B^*) \leq \{V\} = P(A^*,B^*)$. The inequality $P(A^*,B^*) \leq P(A^*,B)$ is proved similarly.

REFERENCES

1. ELLIOTT, R.J. The existence of value in stochastic differential games. S.I.A.M. Jour. Control 14(1976), 85-94.

2. ELLIOTT, R.J. and KALTON, N.J. The existence of value in differential games, Memoir of the American Math. Society 126, Providence, R.I. 1972.

3. ELLIOTT, R.J. and KALTON, N.J. The existence of value in differential games of pursuit and evasion, J. Diff. Equations 12(1972), 504-523.

4. ELLIOTT, R.J. and KALTON, N.J. Cauchy problems for certain Isaacs-Bellman equations and games of survival. Trans. Amer. Soc. 198(1974), 45-72.

5. FRIEDMAN, A., Differential Games, Pure and Applied Math., Vol. 25, Wiley-Interscience, New York 1971.

6. ROXIN, E.O. The axiomatic approach in differential games. J. Opt. Theory and App. 3 (1969), 153-163.

7. VARAIYA, P.P. and LIN, J., Existence of saddle points in differential games. S.I.A.M. Jour. Control, 7(1969), 141-157.

TOWARD A GENERAL THEORY OF PURSUIT AND EVASION

Otomar Hájek
Department of Mathematics and Statistics
Case Western Reserve University
Cleveland, OH 44106/U.S.A.

ABSTRACT

This is a brief introduction to several open problems in the following topics of differential games: rôle determinacy, singular surfaces, feedback strategies, games on manifolds, a new type of problem.

The relatively new subject of differential games has experienced quite explosive growth. The exterior characteristics are conferences, symposia, and workshops (Annapolis, Rhode Island [twice], Warwick, Twente); many papers and reports, and at least four monographs. Internally, the groundwork has been laid, basic concepts and methods identified (at least tentatively), crucial examples assembled and assimilated (at least partially). In stochastic games we have even acquired a step-child. Some workers are retiring from the field, others producing, and new ones entering. Mathematical Reviews has recognised the subject (classification number 90D25), the Encyclopaedia Britannica has not, as yet.

It may be time to strive toward a general theory: not individual cases, however instructive they may be, not intuition, definitely not further generalisations (we have to furnish our empty house first). This is the reason for the title; albeit a more accurate one would be, "Hopefully, more or less in the direction of a possibly general theory of pursuit and evasion." It seems probable (to the author at least: but see the other titles of announced papers here) that progress in the directions listed in the Abstract is crucial to the further development of the theory of pursuit-evasion games, and possibly to all of differential game theory.

Rôle Determinacy and Symmetric Games

In the situation of a torpedo homing in on its target, or of a large airplane evading a possibly dangerous small plane not equipped with radar, the rôles of the two players are determined in advance: one is the pursuer, the other constantly evades. On the other hand, if several destroyers are attempting to neutralize a nimble submarine, carelessness might lead to one of the pursuing ships being, in turn, hunted and sunk; and in an airplane dogfight it is obvious that the pursuer-evader rôle switches back and forth (and even the threat of a decisive rôle determiniation is operative).

The type of game treated in [2] was, roughly, as follows. The state variable x follows

$$\dot{x} = f(x,p,q); \tag{1}$$

the pursuer desires to have it reach a given target, $x(t) \in \Omega$, the quarry wishes to avoid this; the pursuer's strategy is a mapping $\sigma = \sigma(q(\cdot),t)$ depending on $t \geq 0$ and on the behaviour of quarry control functions $q(\cdot)$ only over the time interval $[0,t]$. The rule is that, for any quarry choice $q(\cdot)$, the pursuer control p has $p(t) = \sigma(q(\cdot),t)$.

This information pattern was both amenable and adequate for the applications suggested in [2]. E.g. the Incompatible Strategies Paradox was ruled out by definition: quarry has available only control functions $t \mapsto q(t)$, not strategies. One consequence is that, again, the players' rôles are fixed; another is that they are highly asymmetric , with the evasion problem far harder than that of pursuit.

It is not particularly difficult to describe a more general situation, pursuit-evasion games which form a <u>symmetric class</u>, in the sense that if a game belongs to the class, then so does that obtained by interchanging the pursuer-quarry labels. (A warning: in the evolution of game theory, there never was a shortage of definitions, but only of useful ones. However, mere taxonomy is not mathematics.)

Again describing only roughly, one might be given a dynamical equation as above (with state variable x encompassing the possibly separated phases x_1,x_2 of the two players), and also two target sets Ω_1,Ω_2. The first player controls p (in a manner specified later), and attempts to reach Ω_1 while avoiding Ω_2,

$$x(t) \in \Omega_1, \qquad x(s) \notin \Omega_2 \quad \text{for} \quad 0 \leq s < t;$$

dually for the second. There is considerable flexibility here: cooperation, complete $\Omega_2 \subset \Omega_1$ or partial $\Omega_1 \cap \Omega_2 \neq \emptyset$; stand-off situations $x(t) \notin \Omega_1 \cup \Omega_2$ for all $t \geq 0$; $\Omega_2 = \phi$ as special case; and, state constraints $x(t) \notin \Sigma$ for $t \geq 0$ might be included, either obligatory or automatic (i.e., as in encountering a trap-door or a solid wall).

Two questions have been left open: what is the 'information pattern' used by the players in choosing their controls; and whether any useful results can be obtained in this -- admittedly rather general -- situation. Let us first consider the latter.

A quote from [2, Introduction]: "In treating a pursuit game one may address a series of problems of increasing complexity: decide whether there are any winning positions at all (absence of complete evasion); decide..." This was attacked [loc. cit., Chap. VII] by showing that winning initial positions exist if, and only if, the target set contains 'essential' points (limits of winning positions as termination time bound tends to 0), and then by obtaining workable necessary, and also sufficient, conditions for a point to be essential.

To be specific, in the game with dynamical equation (1), let the pursuer target set be described by $\phi(x) \leq 0$. Set up the 'formal time derivative of ϕ along trajectories',

$$\left(\frac{\partial \phi}{\partial x_1}, \ldots, \frac{\partial \phi}{\partial x_n}\right) \cdot f(x,p,q),$$

at a boundary point x of the target. Then for x to be essential it is necessary that the $\max_q \min_p \leq 0$, and sufficient that the $\min_p \max_q < 0$. (All this under the usual assumptions of smoothness, boundedness, etc.) Thus, if Isaacs' condition min max = max min obtains, this is analogous to the second order conditions for extrema of real functions. What is surprising appears in the simple proof of this (loc. cit., pp. 214-216), that <u>constant</u> controls suffice. Thus the conditions for -- as contrasted with the definition of -- essentiality are remarkably insensitive to many of the fine distinctions in the type of strategy allowed.

Now, it seems most likely that analogous conditions will also apply in the symmetric games described above: 'essential' points in $(\partial\Omega_1)\setminus\Omega_2$ or in $(\partial\Omega_2)\setminus\Omega_1$. The added bonus would be that, at least for this particular problem, one can avoid the first question, finesses in specification of player strategies.

Singular Surfaces

The author feels obliged to express his discomfort and confusion on this topic. Introduced by Isaacs, the singular surfaces have proved to be an effective tool; e.g., the boundary of the set of player's winning positions should, in reasonable cases, be such. But Isaacs also warned us not to place too much reliance on the concept.

Consider the following game with the harmonic oscillator. The scalar dynamical equations are

$$\dot{x} = y + u + v\ , \qquad \dot{y} = -x - u + v,$$

with control constraints $-1 \le u,v \le 1$, and $x \le 0$ as termination condition. A simple analysis (loc. cit., pp. 166-169) yields that all initial positions win, within a time π at most. Unless quarry cooperates, points in the first quadrant are forcedinto the fourth, more-or-less rightward, and then to the negative y-axis (the set of essential points). Now, Isaacs' prescription, of passing a semi-permeable surface through the boundary of the essential set, results in two quarter-circles through the origin, with centres at $(0, \pm 2)$. But what have these to do with the game? If nothing, why?

Obviously both insight and results are needed here; and volunteers might find experience with first-order partial equations useful.

Feedback Controls and Generalized Solutions

Even in a game as rudimentary as pursuit in the plane with simple motion, the first solution that comes to mind involves feedback (i.e., dependent on current state) controls. In many natural problems involving optimization, these controls depend discontinuously on the state variable. Then the equation arising from (1), $\dot{x} = f(x,p(x),q(x))$, may well have the (composed) right-hand side discontinuous. What is then a trajectory solution $x(\cdot)$? A little experimentation readily convinces one that an answer, if there is one at all, is neither easy nor obvious.

By quite different methods, Hermes and Krasovskij arrived at one and the same solution concept, related to earlier work of Filippov. Some terminology is useful to describe this. Given the equation

$$\dot{x} = f(x) \tag{2}$$

with $f : R^n \to R^n$ possibly discontinuous, a function $x : [0,1] \to R^n$ is a Newton solution iff it satisfies (2) for all times $t \in [0,1]$; it is a Carathéodory solution if it is absolutely continuous (A.C.) and satisfies (2) almost everywhere.

Next, an A.C. $x : R^1 \to R^n$ is a Filippov solution [1] iff, a.e.,

$$\dot{x}(t) \in \tilde{f}(x(t)),$$

where

$$\tilde{f}(x) = \bigcap_{\varepsilon>0} \bigcap_{Z} \overline{\text{cvx}}\, f((x + \varepsilon B)\setminus Z)$$

(Z varies over sets of n-dimensional measure 0, $\overline{\text{cvx}}$ denotes the closure of convex hull, and B is the solid unit ball in R^n). If the zero-sets are omitted, one obtains the Krasovskij solutions, [7, pp. 40-41]. Finally, an A.C. $x : [0,1] \to R^n$ is a Hermes solution iff there exist measurable $p_k : R^1 \to R^n$ such that $p_k \to 0$ uniformly (the 'inner perturbations') and Carathéodory solutions x_k of $\dot{y} = f(y + p_k(t))$ such that $x_k \to x$ uniformly. (Obvious modifications if [0,1] is replaced by other intervals, say open or unbounded.) (This last definition is implicit in [5].)

For convenience, let $\mathcal{N}, \mathcal{C}, \mathcal{F}, \mathcal{K}, \mathcal{H}$ denote the corresponding classes of solutions of (2). Under the mild assumption that f is bounded on every ball, obviously

$$\begin{matrix} \mathcal{N} \subset \mathcal{C} \subset \mathcal{H} \\ \quad\quad\ \ \cap \\ \quad \mathcal{F} \subset \mathcal{K} \end{matrix} \tag{3}$$

Some examples are in place. Let $n = 2$, $f = (u,v)$. First, $u = 1$, $v = -1$ in the upper half plane, $u = v = 1$ in the lower, and $u = -1$, $v = 0$ on the x-axis. Then the x-axis traversed negatively is an $\mathcal{N}$-solution which is not $\mathcal{F}$. Second, only change to $u = v = 1$ on the x-axis also. Then the x-axis traversed positively is an $\mathcal{F}$-solution which is not $\mathcal{C}$; in fact, there is no $\mathcal{C}$-solution through points on the x-axis.

One rather deep assertion of [5] may now be re-formulated as $\mathcal{F} \subset \mathcal{H}$. With the proof as model, actually $\mathcal{K} \subset \mathcal{H}$ can be proved, and then a 'closure theorem' yields the converse containment: i.e., the most unexpected

$$\mathcal{K} = \mathcal{H} \ . \tag{4}$$

The same closure theorem, applied to the classical method of Euler broken lines (which are indeed $\mathcal{C}$-solutions of inner perturbations) provides a general existence theorem for $\mathcal{H}$-solutions of (2), without assumptions on f (not even measurability!). Since $\mathcal{H}$-solutions are limits of perturbations, $\mathcal{H}$-uniqueness is equivalent to the formally stronger requirement of continuous dependence on initial data and on inner and outer perturbations. Now, Filippov provided directly verifiable conditions for $\mathcal{F}$-uniqueness, assumptions involving a version of piecewise continuity of f: under like conditions, $\mathcal{F} = \mathcal{H}$ is easily proved. Thus one has available useful conditions for continuous dependence of $\mathcal{F} = \mathcal{K} = \mathcal{H}$ solutions. (The details of this assembling of the results of [1], [5], [7] are in a forthcoming paper, [3].)

Three open questions in this context. Equation (2) is autonomous; and, e.g. for the existence theorem above, non-autonomous equations must be made autonomous by adding time in an extra dimension. Thus the inner perturbations in $\mathcal{H}$, and the ε-balls in $\mathcal{K}$ are to apply to the time-variable also, in the form of infinitesimal time deviations. This is not so in Filippov's original definition. I do not know whether this can also be equivalently carried out for the $\mathcal{H}$ and $\mathcal{K}$ concepts. (If not, then it could no longer be said that a mathematician is one who cannot tell the difference between a signpost and a clock.)

Second, the concept of $\mathcal{F}$-solutions was, obviously, designed so as not to be affected by changes of f in (2) on zero-sets. Does it really do so? Obviously yes, in one direction. However, is it also true that, conversely, if a $\mathcal{H}$-solution x remains an $\mathcal{H}$-solution whenever f is varied on any zero-set, then necessarily x is an $\mathcal{F}$-solution? (If not, there must be another, better, $\mathcal{F}$-solution-like concept.)

Last, from (3) and (4) we have $\mathcal{C} \cup \mathcal{F} \subset \mathcal{H}$ (remark: for suitable equations (2), $\mathcal{C} \cap \mathcal{F} = \phi$ is possible); is this an equality? If not, is there some interesting type $\mathcal{J}$ of solutions such that $\mathcal{C} \cup \mathcal{F} \cup \mathcal{J} = \mathcal{H}$?

Linear Games on Nonlinear Manifolds

Continuing the quotation from the first paragraph, one of the desiderata is, "for each winning position [to] construct a winning strategy..." Of course, this is quite hopeless: one cannot even construct, in any reasonable sense, the solution of an individual differential equation, except in the case of linear equations. Thus, linearity is to be cherished, and nurtured.

The kinematic equation of a player with simple motion in the plane is

$$\dot{x} = u \, , \qquad \|u(t)\| \le 1 \; ; \tag{4*}$$

the equation is linear, with trivial coefficient matrix $A = 0$. Analogously if the plane is replaced by the unit 2-sphere S^2 in R^3,

$$S^2 = \{x \in R^3 : \|x\| = 1\}$$

(see Kelley's game in [2]). The kinematic equation is (4^*) in dimension 3, with the added restriction that $x(t)$ remain on S^2 throughout. Obviously one reformulation of this is that always $u(t)$ be perpendicular to $x(t)$; or that $u(t)$ be in the tangent space at $x(t)$ to the manifold S^2. The control constraint in (4^*) can obviously be interpreted in terms of the natural Riemann metric on S^2 as in (5) below. Now, there is a reason for employing all these important-sounding terms: the kinematic equation itself is still linear!

As a second example, the Game of Two Cars has the following representation (in 'reduced' state space)

$$\dot{x} = yp - s_1 + s_2 \cos\theta,$$

$$\dot{y} = -xp - s_2 \sin\theta,$$

$$\dot{\theta} = p - q,$$

with control constraints

$$|p(t)| \le \frac{s_1}{\rho_1} \, , \qquad |q(t)| \le \frac{s_2}{\rho_2}$$

(the s_i, ρ_i are constant parameters: speed and curvature bound). If one were confronted with this system without explanation, it would take a major effort to discover that the original system, in space of dimension 6, was actually linear (roughly, and in another notation, $\dot{x} = u$, $\dot{y} = v$; the state space is $R^2 \times S^1 \times R^2 \times S^1$). The non-linearity was incurred in lowering the space dimension 6 to a more workable 3.

The immediate generalization of (4^*) is

$$\dot{x} = u \, , \qquad u'G(x)u \le 1 \tag{5}$$

with x on a Riemannian manifold M, u in the tangent space $T_x(M)$, and $G(x) = (g_{ij}(x))$ the matrix of the Riemann metric of M at x. Here, however, the 'coefficient matrix' is still trivial, and one ought to look for versions of (5) with $\dot{x} = Lx + u$ and L "linear" (or rather, affine, in the sense of linear plus constant: there is no distinguished origin in M). There should then be a nice formula for the solution x, possibly involving an 'exponential mapping' associated with L and M. The first snag is L; evidently $L : M \to T(M)$ is a vector field which should be linear or affine in some sense.

One solves, partially, by generalizing: what is an affine map $L : M \to N$ between Riemannian manifolds? My colleagues in differential geometry tell me that a reasonable answer is, the 'totally geodetic maps'; these map every segment of a geodesic in M into a geodesic in N; evidently this does specialize correctly for Euclidean spaces. (E.g., in (5), for $x \in M$ we have $Lx = (x,0) \in T_x(M)$.) There is a second problem: we wish to take $T(M)$ for N (or, referring to $\ddot{x} = u$, even $T^2(M)$). Now $T(M)$ is a differentiable manifold in exactly one natural way, but there appear to be at least five reasonable ways of choosing a Riemann metric [8]. It would make a fascinating project to see which of these, or which other, is particularly appropriate for our control- and game-theoretic purposes.

Conceivably the outlined approach may not be appropriate. Possibly it is far too general: the usual examples involve positions, velocities, and attitudes of $n \geq 2$ players, so that the given state spaces, $R^{3n} \times (S^2)^n$, are quite special; and this might be used to some advantage.

A Numerical Experiment

Consider some simple game, e.g.

$$\ddot{x} + x = u - v \; ; \qquad |u| \le 1 \, , \quad |v| \le \frac{1}{2} \; ; \qquad \text{end:} \quad x = 0 = \dot{x}.$$

A conceptually simple method, possibly not optimal, was suggested in [2]: the player controlling u first neutralizes the opponent,

$$u = v + w \, , \qquad |w| \le \frac{1}{2}$$

and uses left-over capability w to steer to the origin. What would happen if v has a short burst of magnitude not conforming to $|v| \le \frac{1}{2}$, while u is committed to the previous rule (modified by, say, $u = \operatorname{sgn} v$ whenever $|v| > 1$)? Slightly more concretely, let v be semi-periodic with period $2T$, value 0 in $[0,T-\delta]$, value $\alpha > 1$ in $(T-\delta,T)$; how does the state variable $(x,\dot{x})$ behave?

This might make for an interesting student project, with some follow-ups: are there 'bad' values of $2T$? If so, are they connected to the natural frequency in the kinematic equation? Would an asymptotically stable equation improve matters? I confess to being old-fashioned, but I would really like to see the oscilloscope read-out of an analogue realization of this problem.

References

[1] Filippov, A. F., Differential equations with discontinuous right-hand sides (in Russian), Mat. Sbornik, 1960.

[2] Hájek, O., Pursuit Games, Academic Press, 1975.

[3] Hájek, O., Generalized solutions of ODE (to appear).

[4] Hájek, O., Inverted paradoxes in optimal control theory, Preprint no. 275, Fachbereich Mathematik, TH Darmstadt, 1976.

[5] Hermes, H., Discontinuous vector fields and feedback control (in Proc. of Puerto Rico Symposium; eds., J. P. LaSalle and J. Hale), Academic Press, 1967.

[6] Isaacs, R., Differential Games, Wiley, 1967.

[7] Krasovskij, N.N., and Subbotin, A. I., Positional Differential Games (in Russian), Izd. Nauka, 1974.

[8] Yano, K., and Ishihara, S., Differential Geometry of Tangent and Cotangent Bundles Dekker, 1973.

Supplementary reference:

[9] Leitmann, G., A note on optimal open-loop and closed-loop control.

MANY PLAYER DIFFERENTIAL GAMES

G. Leitmann
Department of Mechanical Engineering
University of California
Berkeley, California 94720

1. Introduction

We are concerned with situations in which a number of decision makers - we call them players since we adopt a game setting - are involved. Each player has a cost, payoff or performance criterion, that he wishes to extremize. Each player's cost depends not only on his decision but also on those of the other players. Such situations arise in many ways; eventually we treat some typical examples such as profit maximization through advertising and stabilization in the presence of uncertainty.

To be specific we consider the following setting. Suppose there are N players. The rules of the game assign to each player a decision set and a cost function. Let D_i be the decision set of player i and let his cost function be

$$J_i(\cdot) : D \to R^1 \qquad i = 1, 2, \ldots, N \tag{1.1}$$

where $\quad D \subseteq \prod_{i=1}^{N} D_i \quad$ is also prescribed.

Each player desires to attain the smallest possible cost to himself. If there exists a decision N-tuple $d^u \in D$ such that for all $i \in \{1, 2, \ldots, N\}$

$$J_i(d^u) \leqslant J_i(d) \quad \forall\, d \in D$$

then d^u is a desirable decision N-tuple, for it simultaneously minimizes each player's cost. In general, no such utopian decision exists and the players are faced with a dilemma : What is an optimal decision or, to put it differently, what mood of play should they adopt? Should they cooperate in arriving at a joint decision or not? Various moods of play in a dynamic setting have been explored; e.g., see Refs. 1-6. Here we restrict the discussion to non-cooperative play in the sense of Nash (Ref. 7); that is, each player strives to attain the minimum of his own cost regardless of the consequences to the other players. However, in selecting his "best" decision, what should a player assume about the other players' decisions, for they also affect his cost? Here we consider situations in which each player assumes that his opponents are "rational" and hence make their decisions so as to minimize their costs; that is, each player plays with a view towards minimizing his own cost and not towards "hurting" the other players. We are led to

Definition 1.1 A decision N-tuple $d^e \in D$ is an equilibrium iff for all $i \in \{1, 2, \ldots, N\}$

$$J_i(d^e) \leqslant J_i(d^i)$$

for all $d^i \in D$, where $d^e = \{d_1^e, d_2^e, \ldots, d_N^e\}$ and $d^i = \{d_1^e, \ldots, d_{i-1}^e, d_i, d_{i+1}^e, \ldots, d_N^e\}$.

To illustrate this notion consider a two player game with $D_i \subseteq R^1$ and $D = D_1 \times D_2$, and plot the isocost contours for each player in decision space. The equilibrium, d^e, is a point where the players simultaneously attain minima of their costs. In other words, for each $d_1 \in D_1$ one determines the $d_2 \in D_2$ that corresponds to the smallest value of $J_2(d)$, and conversely for each $d_2 \in D_2$ one determines the $d_1 \in D_1$ that yields the smallest value of $J_1(d)$; of course, these minimizing d_2 and d_1 need not be unique. Each intersection - in general, there can be more than one - of the loci of the minimizing d_2 and d_1 is an equilibrium.

An important class of Nash equilibrium games is that of <u>two-person zero-sum</u> games. In such games there are but two players, and the cost of one player is the negative of the other's; one player loses what the other gains. Thus, one has

$$J_1(d) = - J_2(d) \triangleq J(d)$$

Hence, in terms of $J(d)$, player 1 is the minimizing player and player 2 is the maximizing one. For such a game, Definition 1.1 becomes

<u>Definition 1.2</u> A decision pair $d^e \in D$ is a <u>saddlepoint</u> iff

$$J(d^2) \leqslant J(d^e) \leqslant J(d^1)$$

for all d^1 and $d^2 \in D$, where $d^e = \{d_1^e, d_2^e\}$, $d^1 = \{d_1, d_2^e\}$, $d^2 = \{d_1^e, d_2\}$.

If $D = D_1 \times D_2$ - that is, if the players can make independent decisions - then two-player zero-sum games enjoy some important special properties. Among these are the following (e.g., see Refs. 5,8,9):

(i) $$\inf_{d_1 \in D_1} \sup_{d_2 \in D_2} J(d_1, d_2) \geqslant \sup_{d_2 \in D_2} \inf_{d_1 \in D_1} J(d_1, d_2) .$$

If a saddlepoint exists, then

(ii) $J(d^e)$ is unique and is called the <u>value</u>, J^e ,

(iii) $$\min_{d_1 \in D_1} \sup_{d_2 \in D_2} J(d_1, d_2) = \max_{d_2 \in D_2} \inf_{d_1 \in D_1} J(d_1, d_2) = J^e ,$$

(iv) $$J(d_1^e, d_2) \leqslant \min_{d_1 \in D_1} \sup_{d_2 \in D_2} J(d_1, d_2) \quad \forall\, d_2 \in D_2$$

$$J(d_1, d_2^e) \geqslant \max_{d_2 \in D_2} \inf_{d_1 \in D_1} J(d_1, d_2) \quad \forall\, d_1 \in D_1 .$$

Property (iv) is especially important because, according to it, the adoption of decision d_i^e by player i assures him a cost that is at least as favorable as the best one he can guarantee to himself. However, it must be stressed that these properties depend on the players' ability to make <u>independent</u> decisions; they need not hold otherwise (e.g., see Ref. 10). This point must be recalled later when we discuss the two-player zero-sum version of differential games.

2. Dynamical System

2.1 State Equation

We are concerned with a dynamical system defined by n real numbers, $x \in R^n$, loosely called the state of the system, which change in a prescribed manner with the passing of time $t \in (-\infty, \infty)$; of course, any "time-like" variable may serve as independent variable. The evolution of the state is influenced, or controlled, by N agents, the players. In particular, we deal here with a dynamical system governed by ordinary differential equations.

Given an initial state x^o at time t_o, we are interested in the motion of the state under the players' control. Consider functions

$$u^k(\cdot) : [t_o, t_1] \to R^{s_k} \quad , \quad k = 1, 2, \ldots, N, \quad t_1 > t_o$$

generally taken to be Lebesgue measurable and bounded, and a prescribed C^1 function

$$f(\cdot) : R^n \times R^1 \times R^{s_1} \times \ldots \times R^{s_N} \to R^n$$

The state equation is

$$\dot{x}(t) = f(x(t), t, u^1(t), \ldots, u^N(t)) \tag{2.1}$$

For given control variables $u^k(\cdot)$ and initial state x^o, the solution of (2.1) is an absolutely continuous function

$$x(\cdot) : [t_o, t_2] \to R^n, \; x(t_o) = x^o \; , \; t_o < t_2 \leqslant t_1 \; .$$

2.2 Strategies

The players influence the evolution of the state through their choices of the $u^k(t)$ for almost all $t \in [t_o, t_1]$. This raises the question : What information is available to the players, that is, on what can they base their choices? They might have "no information", in which case they are forced to play open-loop; in that event, each player selects a function of time on a bounded interval as his control. Since all players influence the state evolution, knowledge of the current state gives each player some information about the other players' past actions. Hence we consider a somewhat more reasonable situation, namely, one in which each player has access to complete state information; that is, at almost every instant of time, t, each player knows the exact value of the state, x. A player's control then is a function from state-time space into his control space; we call such a function a feedback or closed-loop control, or simply a strategy[†]. It must be noted that the question of information is an important one which we do not explore here; for a treatment of this question see Ref. 13-15, among others.

A strategy for player k is a function

$$p^k(\cdot) : R^n \times R^1 \to R^{s_k}$$

Then, if $x(t)$ is the state at time t,

[†]There are other definitions of strategy; e.g., see Refs. 11-12.

$$u^k(t) = p^k(x(t), t)$$

With this definition of strategy the dynamical system becomes a generalized one; e.g., see Refs. 16 and 17. Thus, in place of state equation (2.1) we have

$$\dot{x}(t) \in K(x(t), t) \quad , \quad x(t_0) = x^0 \tag{2.2}$$

where, for given $p^k(\cdot)$, $k = 1, 2, \ldots, N$,

$$K(x, t) \triangleq \{z \in R^n \mid z = f(x, t, p^1(x, t), \ldots, p^N(x, t))\}$$

Since we are not concerned here with existence of solutions - we consider this question subsequently in the application to stability - we need not pursue this topic with the exception of allowing for nonuniqueness of solutions.

However, we place certain restrictions on the sets of <u>admissible strategies</u>; such restrictions arise in part from the problems under consideration and in part from the requirements of the mathematical treatment. The former are in the form of constraints. For $k \in \{1, 2, \ldots, N\}$ let

$$U^k(\cdot) : R^n \times R^1 \to \text{set of all nonempty subsets of } R^{s_k}$$

be prescribed set-valued functions; then, given $(x, t) \in R^n \times R^1$, $U^k(x, t)$ is the set of all strategy values available to player k at state x and time t .

Furthermore, given a strategy N-tuple $p(\cdot) = \{p^1(\cdot), \ldots, p^N(\cdot)\}$ and initial state x^0 , let $\mathcal{U}^k(x^0, t_0)$ denote the set of all control functions, $u^k(\cdot)$, corresponding to all solutions, $x(\cdot)$, of (2.2), where $u^k(t) = p^k(x(t), t)$.

Now we state

<u>Definition 2.1</u> Set P is a set of admissible strategy N-tuples iff for all $p(\cdot) \in P$ and $(x^0, t_0) \in R^n \times R^1$

(i) $u^k(\cdot) \in \mathcal{U}^k(x^0, t_0)$, $k = 1, 2, \ldots, N$, is Lebesgue measurable and bounded;

(ii)† $p^k(x, t) \in U^k(x, t) \quad (x, t) \in R^n \times R^1$;

(iii) if $\bar{p}(\cdot)$, $\bar{\bar{p}}(\cdot) \in P$ and $\bar{t} \in (-\infty, \infty)$, and $p(\cdot)$ is such that

$$p(x, t) = \bar{p}(x, t) \quad \text{for} \quad t \leq \bar{t}$$
$$p(x, t) = \bar{\bar{p}}(x, t) \quad \text{for} \quad t > \bar{t} ,$$

then $p(\cdot) \in P$.

Condition (i) assures that the integrals in the costs to be defined subsequently do exist. Condition (ii) imposes prescribed constraints on the strategy values. Condition (iii) is a technical restriction that permits the "joining" of strategies along the time axis; it is imposed to validate a "principle of optimality" used in the derivation of necessary conditions.

2.3 <u>Playability</u>

We suppose that, among their motivations, all players share the desire to steer the

† It is also possible to impose state constraints, although we do not do so here.

state from a given initial one, (x^0, t_0), to one belonging to a prescribed <u>target</u> set $\theta \subset R^n \times R^1$; that is, having begun play at state x^0 and time t_0 , play ends when $(x, t) \in \theta$. Of course, there are situations, such as pursuit-evasion ones, in which one player desires reaching a target while the other wishes to avoid it. Such games are basically of another type called "games of kind" or "qualitative games" in counterdistinction to the "games of degree" or "quantitative games" considered here; for a discussion of the former, see Refs. 18-20, among others. If termination is not desired by all players, an alternative to playability is proposed in Ref. 21. Now we introduce

<u>Definition 2.2</u> A strategy N-tuple $p(\cdot)$ is <u>playable</u> at (x^0, t_0) iff it is admissible and generates at least one solution, $x(\cdot)$, of (2.2) such that $x(t_0) = x^0$, $(x(t), t) \notin \theta$ for $t \in [t_0, t_1)$, and $(x(t_1), t_1) \in \theta$. Such a solution is called <u>terminating</u> and t_1 is the corresponding <u>terminal time</u>. A quadruple $\{x^0, t_0, p(\cdot), x(\cdot)\}$, where $x(\cdot)$ is a solution generated by $p(\cdot) \in P$ and $x(t_0) = x^0$, is termed a <u>play</u>. A play is a terminating one iff $x(\cdot)$ is terminating.

Note that Definition 2.2 embodies the shared desire for termination on the given target set and that termination occurs the first time $(x, t) \in \theta$. Subsequently we shall have more to say about the consequences of imposing playability. Here we call attention to the fact that the i-th players decision, d_i , corresponds to a quadruple $\{x^0, t_0, p^i(\cdot), x(\cdot)\}$ and set D to the set of all terminating plays $\{x^0, t_0, p(\cdot), x(\cdot)\}$; hence, in general, decisions cannot be made independently, since solution $x(\cdot)$ depends on the strategies of all players and, furthermore, need not be unique.

2.4 <u>Costs</u>

Before defining a mood of play, in this case Nash equilibrium, we must introduce the players' cost functions. Let C^a be the class of absolutely continuous functions from $R^1 \to R^n$. Then consider functions

$$V_i(\cdot) : R^n \times R^1 \times P \times C^a \to R^1$$

such that

$$V_i(x^0, t_0, p(\cdot), x(\cdot)) = \int_{t_0}^{t_1} f_0^i(x(t), t, u^1(t), \ldots, u^N(t))\, dt \tag{2.3}$$

where $f_0^i(\cdot) : R^n \times R^1 \times R^{s_1} \times \ldots \times R^{s_N} \times R^1$ are given C^1 functions and

$u^i(t) = p^i(x(t), t)$.

Function $V_i(\cdot)$ so defined is the cost function of player i ; it depends only on play $\{x^0, t_0, p(\cdot), x(\cdot)\}$.

2.5 Nash Equilibrium

Now we are ready to restate Definitions 1.1 and 1.2 for the class of differential games considered here.

Definition 2.3 A strategy N-tuple ${}^{e}p(\cdot)$ is an equilibrium on a set $X \subseteq R^n \times R^1$ iff

(i) it is playable at all $(x^o, t_o) \in X$,

and for all $i \in \{1, 2, \ldots, N\}$ and $(x^o, t_o) \in X$

(ii) $V_i(x^o, t_o, {}^{e}p(\cdot), x^e(\cdot)) \leq V_i(x^o, t_o, {}^{i}p(\cdot), x^i(\cdot))$

for all terminating plays $\{x^o, t_o, {}^{e}p(\cdot), x^e(\cdot)\}$ and $\{x^o, t_o, {}^{i}p(\cdot), x^i(\cdot)\}$, where ${}^{i}p(\cdot) = \{{}^{e}p^1(\cdot), \ldots, {}^{e}p^{i-1}(\cdot), p^i(\cdot), {}^{e}p^{i+1}(\cdot), \ldots, {}^{e}p^N(\cdot)\}$.

For two-person zero-sum games, let

$$V_1(\cdot) = - V_2(\cdot) = V(\cdot)$$

so that Definition 2.3 becomes

Definition 2.4 A strategy pair ${}^{e}p(\cdot)$ is a saddlepoint on a set $X \subseteq R^n \times R^1$ iff

(i) it is playable at all $(x^o, t_o) \in X$,

and for all $(x^o, t_o) \in X$

(ii) $V(x^o, t_o, {}^{2}p(\cdot), x^2(\cdot)) \leq V(x^o, t_o, {}^{e}p(\cdot), x^e(\cdot)) \leq V(x^o, t_o, {}^{1}p(\cdot), x^1(\cdot))$

for all terminating plays $\{x^o, t_o, {}^{2}p(\cdot), x^2(\cdot)\}$, $\{x^o, t_o, {}^{e}p(\cdot), x^e(\cdot)\}$ and $\{x^o, t_o, {}^{1}p(\cdot), x^1(\cdot)\}$, where ${}^{1}p(\cdot) = \{p^1(\cdot), {}^{e}p^2(\cdot)\}$ and ${}^{2}p(\cdot) = \{{}^{e}p^1(\cdot), p^2(\cdot)\}$.

Some observations are in order. First of all, it is readily established that non-uniqueness of solutions does not affect the equilibrium values of the costs. Thus, if a given equilibrium ${}^{e}p(\cdot)$ generates two terminating solutions, say $x^e(\cdot)$ and $\bar{x}^e(\cdot)$, then

$$V_i(x^o, t_o, {}^{e}p(\cdot), x^e(\cdot)) = V_i(x^o, t_o, {}^{e}p(\cdot), \bar{x}^e(\cdot))$$

Furthermore, in view of (2.3) and the definition of playability, Definition 2.2,

$$V_i(x^o, t_o, {}^{e}p(\cdot), x^e(\cdot)) = 0 \quad \forall\ (x^o, t_o) \in \theta$$

Thus, given ${}^{e}p(\cdot)$, one can define functions

$$V_i^e(\cdot) : X \cup \theta \to R^1 \tag{2.4}$$

such that for all $(x^o, t_o) \in X \cup \theta$

$$V_i^e(x^o, t_o) = V_i(x^o, t_o, {}^{e}p(\cdot), x^e(\cdot))$$

for all terminating plays $\{x^o, t_o, {}^{e}p(\cdot), x^e(\cdot)\}$.

Of course, for two-person zero-sum games,

$$V_1^e(\cdot) = -V_2^e(\cdot) \triangleq V^e(\cdot)$$

and $V^e(\cdot)$ is the value function.

However, if equilibrium is not unique -- and often it is not -- the players' equilibrium costs may differ. That is, suppose ${}^e p(\cdot)$ and ${}^e \bar{p}(\cdot)$ are equilibria, and $V_i^e(\cdot)$ and $\bar{V}_i^e(\cdot)$ are the corresponding cost functions (2.4); then, in general

$$V_i^e(x^o, t_o) \neq \bar{V}_i^e(x^o, t_o)$$

However, for two-person zero-sum games some properties follow readily from Definition 2.4. Among these (e.g., see Ref. 4) are the following.

Lemma 2.1 Suppose that ${}^e p(\cdot)$ and ${}^e \bar{p}(\cdot)$ are saddlepoints on X. If, given $(x^o, t_o) \in X$, $\{x^o, t_o\ {}^e p(\cdot), x^e(\cdot)\}$ and $\{x^o, t_o, {}^e \bar{p}(\cdot), \bar{x}^e(\cdot)\}$ are terminating plays, and both $\{{}^e p^1(\cdot), {}^e \bar{p}^2(\cdot)\}$ and $\{{}^e \bar{p}^1(\cdot), {}^e p^2(\cdot)\}$ are playable at (x^o, t_o), then

$$V(x^o, t_o, {}^e p(\cdot), x^e(\cdot)) = V(x^o, t_o, {}^e \bar{p}(\cdot), \bar{x}^e(\cdot))$$

For instance, if every admissible strategy pair is playable at $(x^o, t_o) \in X$, the conditions of the lemma are met.

Lemma 2.2 Suppose that ${}^e p(\cdot)$ and ${}^e \bar{p}(\cdot)$ are saddlepoints on X. If both $\{{}^e p^1(\cdot), {}^e \bar{p}^2(\cdot)\}$ and $\{{}^e \bar{p}^1(\cdot), {}^e p^2(\cdot)\}$ are playable at all $(x^o, t_o) \in X$, then they are also saddlepoints on X.

Thus, if the conditions of the lemma are met, the combination of any saddlepoint strategy for player 1 with any saddlepoint strategy for player 2 results in a saddlepoint.

In general, not all strategy pairs are playable; that is, the players cannot assure playability by independent choice of their strategies. Consequently, in a given situation there may be many saddlepoints, each with a different value function. Then the special properties enjoyed by two-person zero-sum games with independent decisions need no longer hold; e.g., see Refs. 22 and 23.

Now we turn to conditions that must be met by an equilibrium strategy N-tuple, that is, necessary conditions.

2.6 Necessary Conditions for Equilibrium

Under certain restrictions one can derive conditions that must be satisfied by an equilibrium strategy N-tuple. The necessary conditions given here are based on geometric arguments first used for two-person zero-sum games (e.g., see Refs. 19 and 24) and subsequently for nonzero-sum games (Ref. 4). Other treatments of zero-sum games are in Refs. 11, 12, 18, 25 and 26,among others; nonzero-sum games are also treated in Ref. 27.

Before presenting a set of necessary conditions, we require a definition.

Definition 2.5 Let $x^e(\cdot) : [t_o, t_1^e] \to R^n$ be a terminating solution generated by equilibrium N-tuple ${}^e p(\cdot)$. This solution is regular iff

(i) $(x^e(t), t) \in \text{int } X \quad \forall\, t \in [t_o, t_1^e)$;

(ii) every $(x^e(t), t)$, $t \in [t_o, t_1^e)$, possesses a neighborhood in $R^n \times R^1$ on which the $V_i^e(\cdot)$ functions are of class C^2 ;

(iii) every $(x^e(t), t)$, $t \in [t_o, t_1^e]$, possesses a neighborhood in $R^n \times R^1$ on which strategies ${}^e p^i(\cdot)$ are of class C^1 ;

(iv) for all $(x^e(t), t)$, $t \in [t_o, t_1^e]$, and all $u^i \in U^i(x^e(t), t)$, there is a strategy $p^i(\cdot)$ of class C^1 on a neighborhood of $(x^e(t), t)$ in $R^n \times R^1$ such that $p^i(x^e(t), t) = u^i$ and $\{{}^e p^1(\cdot), \ldots, {}^e p^{i-1}(\cdot), p^i(\cdot), {}^e p^{i+1}(\cdot), \ldots, {}^e p^N(\cdot)\}$ is admissible.

Furthermore we restrict ourselves to target sets, θ, which are closed in $R^n \times R^1$ and whose boundary, $\partial\theta$, is a smooth manifold; that is, $\partial\theta$ is the intersection of $q \leq n + 1$ smooth surfaces defined by

$$\theta_i(x, t) = 0 \quad , \quad i = 1, 2, \ldots, q \tag{2.5}$$

where

$$\theta_i(\cdot) : R^n \times R^1 \to R^1$$

is of class C^1 and matrix

$$\left[\frac{\partial\theta_i(y)}{\partial y_j} \right] \quad \begin{matrix} i = 1, 2, \ldots, q \\ j = 1, 2, \ldots, n + 1 \end{matrix}$$

has maximum rank for all $y = (x, t) \in \partial\theta$.

In other words, there are q terminal boundary conditions given by (2.5); in the usual situation, there is but one such condition, $q = 1$.

Next let

$$H^i(\lambda^i, x, t, u) \triangleq f_o^i(x, t, u) + \sum_{j=1}^{n} \lambda_j^i f_j(x, t, u) + \lambda_{n+1}^i$$

where $\lambda^i = (\lambda_1^i, \lambda_2^i, \ldots, \lambda_{n+1}^i)^T$ and $u = \{u^1, u^2, \ldots, u^N\}$,

and consider absolutely continuous functions

$$\lambda_j^i(\cdot) : [t_o, t_1^e] \to R^1$$

satisfying

$$\dot\lambda_j^i(t) = -\frac{\partial H^i(\lambda^i, x, t, u)}{\partial y_j} - \sum_{k=1}^{K} \frac{\partial H^i(\lambda^i, x, t, u)}{\partial u_k} \frac{\partial\, {}^e p(x, t)}{\partial y_j} \tag{2.6}$$

$$j = 1, 2, \ldots, n + 1$$

where $y = (x, t)$, $\lambda^i = \lambda^i(t)$, $x = x^e(t)$, $u = p^e(x^e(t), t)$, and $K = s_1 + s_2 + \dots + s_N$.

Then we have

Theorem 2.1 If ${}^e p(\cdot)$ is an equilibrium on X, generating a regular solution $x^e(\cdot) : [t_o, t_1^e] \to R^n$, then for $i \in \{1, 2, \dots, N\}$ there exists a solution $\lambda^i(\cdot) : [t_o, t_1^e] \to R^{n+1}$ of (2.6) such that

(i)
$$\min_{u^i \in U^i(x^e(t), t)} H^i(\lambda^i(t), x^e(t), t, {}^i p(x^e(t), t))$$
$$= H^i(\lambda^i(t), x^e(t), t, {}^e p(x^e(t), t))$$
$$= 0$$

for all $t \in [t_o, t_1^e]$, where

$${}^i p(x, t) = \{{}^e p^1(x, t), \dots, {}^e p^{i-1}(x, t), u^i, {}^e p^{i+1}(x, t), \dots, {}^e p^N(x, t)\},$$

(ii) $\lambda^i(t_1^e)$ is normal to the tangent plane of $\partial\theta$ at $(x^e(t_1^e), t_1^e)$.

If we let $\lambda = \lambda^i$ and $H = H^i$, then for two-person zero-sum games Theorem 2.1 reduces to

Theorem 2.2 If ${}^e p(\cdot) = \{{}^e p^1(\cdot), {}^e p^2(\cdot)\}$ is a saddlepoint on X, generating a regular solution $x^e(\cdot) : [t_o, t_1^e] \to R^n$, then there exists a solution $\lambda(\cdot) : [t_o, t_1^e] \to R^{n+1}$ of (2.6) such that

(i)
$$\min_{u^1 \in U^1(x^e(t), t)} H(\lambda(t), x^e(t), t, u^1, {}^e p^2(x^e(t), t))$$
$$= \max_{u^2 \in U^2(x^e(t), t)} H(\lambda(t), x^e(t), t, {}^e p^1(x^e(t), t), u^2)$$
$$= 0$$

for all $t \in [t_o, t_1^e]$,

(ii) $\lambda(t_1^e)$ is normal to the tangent plane of $\partial\theta$ at $(x^e(t_1), t_1^e)$.

Because of the regularity assumption, these theorems apply to a limited class of problems. Their scope can be enlarged by introducing a decomposition of X and imposing piecewise regularity with respect to that composition; for zero-sum games this is done in Refs. 19, 24, 28 and 29, and for nonzero-sum games in Ref. 4. Such a generalization allows for "singular" surfaces of various types, that is, surfaces on which the equilibrium strategies or their derivatives experience discontinuities; for a discussion of singular surfaces, see Refs. 18, 28, 29 and 30, among others.

Here it should be noted that the necessary conditions embodied in Theorems 2.1 and

2.2 are similar to those for optimal control. In particular, for zero-sum games, Theorem 2.2 is akin to Pontryagin's Maximum Principle (e.g., see Ref. 31). However, there are some important differences. Of course, the Maximum Principle applies without restrictions of regularity; however, it deals with open-loop controls rather than closed-loop ones. This latter difference accounts for the presence of additional terms in equations (2.6), terms not present in the adjoint or costate equations of optimal control theory; these are the terms involving the derivatives of the equilibrium strategies. In the case of zero-sum games and constraint sets $U^i(x, t)$, $i = 1, 2$, defined by inequalities satisfying certain conditions, it can be shown that the additional terms in equations (2.6) vanish; e.g., see Refs. 4 and 19.

Since Theorems 2.1 and 2.2 are only necessary for an equilibrium (in fact, for a subclass of equilibria), it is useful to have conditions whose satisfaction assures an equilibrium.

2.7 Sufficient Conditions for Equilibrium

Here we give the simplest versions of sufficiency conditions for a Nash equilibrium. These conditions fall into one of two classes, "field" conditions and "direct" conditions. An example of the former is

Theorem 2.3 A strategy N-tuple ${}^e p(\cdot)$ that is playable at all $(x^o, t_o) \in X$ is an equilibrium on X with respect to every strategy N-tuple that is playable at (x^o, t_o) and whose terminating solutions $x(\cdot) : [t_o, t_1] \to R^n$ are such that $x(t) \in X \cup \theta$ for all $t \in [t_o, t_1]^\dagger$, if for $i \in \{1, 2, \ldots, N\}$ there exists a C^1 function $V^i(\cdot) : S \to R^1$, where $X \cup \theta \subseteq S$ (open), such that for all $(x^o, t_o) \in X$

(i) $\displaystyle\int_{t_o}^{t_1^e} f_o^i\,(x^e(t), t, {}^e p(x^e(t), t))\,dt = V^i(x^o, t_o)$

for all terminating solutions $x^e(\cdot) : [t_o, t_1^e] \to R^n$ generated by ${}^e p(\cdot)$,

(ii)†† $f_o^i(x, t, {}^i p(x, t)) + \text{grad}\, V^i(x, t)\, f(x, t, {}^i p(x, t)) \geqslant 0$

for all $(x, t) \in X$, $u^i \in U^i(x, t)$, and

(iii) $V^i(x, t) = 0$ for all $(x, t) \in \theta$.

For zero-sum games one can state this theorem involving a single C^1 function $V(\cdot) : S \to R^1$. In that case condition (ii) becomes

$$f_o(x, t, u^1, {}^e p^2(x, t)) + \text{grad}\, V(x, t)\, f(x, t, u^1, {}^e p^2(x, t)) \geqslant 0$$

$$f_o(x, t, {}^e p^1(x, t), u^2) + \text{grad}\, V(x, t)\, f(x, t, {}^e p^1(x, t), u^2) \leqslant 0$$

for all $(x, t) \in X$, $u^i \in U^i(x, t)$, $i = 1, 2$.

---

†That is, ${}^e p(\cdot)$ is compared only to those playable strategy N-tuples which generate terminating solutions that remain in $X \cup \theta$.

††${}^i p(x, t)$ is defined as in Theorem 2.1.

Various versions of this theorem for two-person zero-sum games may be found in Refs. 18, 19, 25 and 28, among others. As in the case of necessary conditions, the introduction of a decomposition of X permits a generalization that enhances the usefulness of the theorem; such a generalization for the nonzero-sum case is in Ref. 32.

For games with fixed time of termination, one can give another kind of sufficiency theorem. Fixed terminal time, t_1, implies that $t = t_1$ if $(x, t) \in \theta$. Now let $\psi^i(\cdot) : [t_0, t_1] \to R^n$ denote absolutely continuous functions. Then one has (e.g., see Ref. 32)

Theorem 2.4 A strategy N-tuple ${}^e p(\cdot)$ that is playable at all $(x^0, t_0) \in X$ is an equilibrium on X with respect to every strategy N-tuple that is playable at (x^0, t_0) and whose terminating solutions $x(\cdot) : [t_0, t_1] \to R^n$ are such that $x(t) \in X \cup \theta$ for all $t \in [t_0, t_1]$, if for $(x^0, t_0) \in X$ and $i \in \{1, 2, \ldots, N\}$

$$\text{(i)} \quad \int_{t_0}^{t_1} f_0^i\,(x^e(t), t, {}^e p(x^e(t), t))\, dt = \text{constant}$$

for all terminating plays $\{x^0, t_0, {}^e p(\cdot)\, x^e(\cdot)$, and there exist $\psi^i(\cdot)$ and a terminating solution $x^e(\cdot)$ generated by ${}^e p(\cdot)$ such that

$$\text{(ii)}^{\dagger} \quad \begin{aligned} & f_0^i(x^e(t), t, {}^e p(x^e(t), t)) + \psi^{iT}(t)\, f(x^e(t), t, {}^e p(x^e(t), t)) \\ & - f_0(x, t, {}^i p(x, t)) - \psi^{iT}(t)\, f(x, t, {}^i p(x, t)) \\ & + \dot{\psi}^{iT}(t)\, [x^e(t) - x] \leqslant 0 \end{aligned}$$

for all $(x, t) \in X$ and all $u^i \in U^i(x, t)$, and

$$\text{(iii)} \quad \psi^{iT}(t_1)\, [x^e(t_1) - x] \geqslant 0 \quad \text{for all } (x, t) \in \theta .$$

The zero-sum version of Theorem 2.4 can be stated involving only one function $\psi(\cdot) = \psi^i(\cdot)$, $i = 1, 2$. In that case, conditions (ii) are given with $\leqslant$ for $i = 1$ and $\geqslant$ for $i = 2$, and condition (iii) applies with equality replacing $\geqslant$.

It is often possible to utilize a solution of equations (2.6) for the test functions $\psi_j^i(\cdot)$, $j = 1, 2, \ldots, n$, such that condition (ii) of Theorem 2.1 is met; that is, $\psi_j^i(\cdot) = \lambda_j^i(\cdot)$.

Now we turn to some applications. We recall that the primary utility of necessary conditions, such as Theorems 2.1 and 2.2, lies in their yielding candidates for equilibria which can then be tested by means of sufficiency conditions. As we noted earlier, this requires solutions of equations (2.6); however, especially because of

---

† ${}^i p(x, t)$ is defined as in Theorem 2.1. Also, condition (ii) need be met only a.e. in $[t_0, t_1]$.

the terms involving derivatives of potential equilibrium strategies, ${}^{e}p(\cdot)$, this is usually impractical. If ${}^{e}p(\cdot)$ is constant with respect to x , that is, depends only on t , then these troublesome terms vanish. For certain classes of problems it is possible to find equilibria which depend only on time, t ; for instance, so-called "trilinear games", treated in Ref. 33, are of this type. Another class of such games is exemplified by the following example.

3. Profit Maximization Through Advertising

A problem of competition between two firms which sell substitutable products is considered in Ref. 34. Each firm desires to maximize its own profits. A firm's share of the market depends on its advertising as well as on that of its competitor. In particular, for $i \in \{1, 2\}$ let

$x_i(t)$ = fraction of total market buying from firm i at time t ,

$u^i(t)$ = rate of advertising expenditure of firm i at time t ,

β_i, c_i k_i = positive constants.

Since a firm's resources are limited, its rate of advertising expenditure is restricted by

$$0 \leq u^i(t) \leq \frac{1}{c_i} \quad , \quad i = 1, 2 \tag{3.1}$$

Also, since the $x_i(t)$ are market fractions

$$\begin{aligned} &0 \leq x_i(t) \quad , \quad i = 1, 2 \\ &x_1(t) + x_2(t) \leq 1 \end{aligned} \tag{3.2}$$

The situation is modelled by

$$\begin{aligned} \dot{x}_1(t) &= -\beta_1\, x_1(t) + u^1(t) - \tfrac{1}{2}\, c_1\, [u^1(t)]^2 - k_1\, x_1(t)\, u^2(t) \\ \dot{x}_2(t) &= -\beta_2\, x_2(t) + u^2(t) - \tfrac{1}{2}\, c_2\, [u^2(t)]^2 - k_2\, x_2(t)\, u^1(t) \end{aligned} \tag{3.3}$$

with $x_i(t_o) = x_i^o$, and fixed horizon (terminal time) t_1 .

Let q_i be the revenue rate of firm i if the entire market buys its product. For sufficiently small horizon, q_i is constant and interest rate is neglected. Thus, the profits over $[t_o, t_1]$ are given by

$$\int_{t_o}^{t_1} [q_i\, x_i(t) - u^i(t)]\, dt \quad , \quad i = 1, 2 \tag{3.4}$$

which firm i seeks to maximize† by appropriate choice of $u^i(t)$, $t \in [t_o, t_1]$. A few words concerning the model are in order. The rate of market share is decreased

† Recall here that necessary and sufficient conditions are given for minimization.

by $\beta_i\ x_i(t)$, that is, firm i's market share decreases without advertising, but is increased by advertising $u^i(t) - \frac{1}{2} c_i\ [u^i(t)]^2$, albeit subject to diminishing returns. Upper bound $\frac{1}{c_i}$ in constraint (3.1) is chosen because it maximizes $u^i - \frac{1}{2} c_i (u_i)^2$; an increase in u^i beyond this value increases advertising cost without yielding an increase in market share rate. The term $-k_i\ x_i(t)\ u^j(t)$ accounts for the decrease of firm i's share due to firm j's advertising which increases in effectiveness with firm i's market share.

Under some reasonable assumptions, state constraints (3.2) are redundant; that is every solution of (3.3) with (3.1) satisfies (3.2). In particular, if initial shares $x_i(t_o)$, $i = 1, 2,$ satisfy (3.2) and

$$\beta_i \geqslant \frac{1}{2c_1} + \frac{1}{2c_2} \quad , \qquad i = 1, 2$$

then (3.2) are redundant.

Now one can show that there exists an equilibrium strategy couple ${}^e p(\cdot) = \{{}^e p^1(\cdot)\ ,\ {}^e p^2(\cdot)\}$ which depends only on time, namely,

$${}^e p^i(x, t) = L_i(t) \quad , \qquad i = 1, 2$$

with

$$L_1(t) = \begin{cases} \dfrac{\lambda_1^1(t) - 1}{c_1\ \lambda_1^1(t)} & \text{if} \quad \lambda_1^1(t) \geqslant 1 \\ 0 & \text{if} \quad \lambda_1^1(t) < 1 \end{cases}$$

$$L_2(t) = \begin{cases} \dfrac{\lambda_2^2(t) - 1}{c_2\ \lambda_2^2(t)} & \text{if} \quad \lambda_2^2(t) \geqslant 1 \\ 0 & \text{if} \quad \lambda_2^2(t) < 1 \end{cases}$$

where for $i \in \{1, 2\}$ the $\lambda_j^i(\cdot)$, $j = 1, 2,$ are solutions of (2.6) subject to condition (ii) of Theorem 2.1; in particular,

$$\dot\lambda_1^1(t) = -\ q_1 + \beta_1\ \lambda_1^1(t) + k_1\ \lambda_1^1(t)\ L_2(t)$$

$$\dot\lambda_2^2(t) = -\ q_2 + \beta_2\ \lambda_2^2(t) + k_2\ \lambda_2^2(t)\ L_1(t)$$

with $\lambda_1^1(t_1) = \lambda_2^2(t_1) = 0$.

Upon use of Theorem 2.4 one can establish readily that ${}^e p(\cdot)$ is indeed an equilibrium.

One can show also that both firms' advertising expenditure rates are zero on some terminal interval; if $\hat t$ is the closest time to t_1 for which

$$\frac{q_i}{\beta_i}\,[1 - \exp \beta_i(\hat{t} - t_1)] = 1$$

for $i = 1$ or 2 , then $L_1(t) = L_2(t) = 0$ for $t \in [\hat{t}, t_1]$.

The problem discussed here is treated in Ref. 34 also for moods of play other than Nash equilibrium.

To recapitulate, we are able to utilize the necessary conditions of Theorem 2.1 since we can determine an equilibrium strategy which depends on time only. Another such class is that of the "trilinear games" of Refs. 33; these are exemplified by some microeconomic applications in Ref. 35.

Sometimes one can deduce equilibria by inspection; examples of such problems may be found in the bargaining situations treated in Ref. 36-38.

A rather different kind of problem is one in which game theory is employed as a tool, for instance to deduce stabilizing controls in the presence of uncertainty. This involves a pseudo-game in that one of the players, Nature, is not a real player, that is, a sentient opponent. Below we give a brief discussion of the simplest problem of this type.

4. Stability in the Presence of Uncertainty

Consider a linear dynamical system

$$\begin{aligned} \dot{x}(t) &= A\,x(t) + B\,u(t) + B\,v(t) \\ x(t_0) &= x^0 \;, \quad t \in R^1_+ \end{aligned} \tag{4.1}$$

where $x(t) \in R^n$, $u(t)$ and $v(t) \in R^p$, and A and B are constant matrices of appropriate dimensions. Here, $u(t)$ is the value of input (control) and $v(t)$ that of input disturbance at time t . We suppose that nothing (of a statistical nature) is known about the input disturbance except that it is bounded; for instance,[†]

$$\| v(t) \| \leq \rho \qquad \forall\; t \in R^1_+$$

where ρ is constant.

We seek a feedback control

$$\tilde{p}(\cdot) : R^n \times R^1_+ \to R^p$$

with

$$u(t) = \tilde{p}(x(t), t)$$

such that origin $\{0\}$ is uniformly asymptotically stable in the large (Lyapunov) no matter what $v(t)$, $t \in R^1_+$.

[†] $\|\cdot\|$ denotes Euclidean norm.

In view of our requirement to guarantee stability under all circumstances, including the worst possible ones, we are led to adopt the philosophy of "worst case" design. Thus, we pose a pseudo-game with Nature as opponent controlling $v(t)$ and seeking to destabilize the system. That is, we consider a two-person zero-sum game for system (4.1) with $u^1 = u$ and $u^2 = v$, and subject to constraints

$$\|u^i(t)\| \leqslant \rho , \quad i = 1, 2 \quad , \quad t \in (-\infty, t_1]$$

Since we desire stability of $\{0\}$, we take a measure of deviation from $\{0\}$; namely,

$$V = \int_{t_0}^{t_1} x^T(t)\, Q\, x(t)\, dt$$

where Q is a constant symmetric $n \times n$ matrix.

Employing necessary conditions, Theorem 2.2, and sufficiency conditions, Theorem 2.3, one finds a saddlepoint given by

$$^e p^1(x, t) = - {}^e p^2(x, t) = - \frac{B^T P(t)\, x}{\|B^T P(t)\, x\|} \rho \quad \text{for} \quad (x, t) \notin N$$

$$^e p^i(x, t) = \text{any admissible value for} \quad (x, t) \in N$$

where

$$N \triangleq \{(x, t) \mid B^T P(t)\, x = 0\}$$

and $P(\cdot)$ is the solution of

$$\dot{P}(t) + P(t)\, A + A^T P(t) + Q = 0$$
$$P(t_1) = 0$$

Having deduced saddlepoint $^e p(\cdot)$ we utilize controller strategy $^e p^1(\cdot)$ as a candidate for a stabilizing control with $t_1 \to \infty$.

With the additional assumptions that Q is positive definite and $p^2(\cdot)$, the disturbance, is continuous in x and piecewise continuous in t, one can show readily that

$$\tilde{p}(x, t) = - \frac{B^T P\, x}{\|B^T P\, x\|} \rho \quad \text{for} \quad (x, t) \notin N$$

$$\tilde{p}(x, t) = u \quad , \quad \|u\| \leqslant \rho , \text{ for } (x, t) \in N$$

where

$$N \triangleq \{(x, t) \mid B^T P\, x = 0\}$$

and P = constant such that

$$P\, A + A^T P + Q = 0 .$$

The details of the proof are in Ref. 38-41. Here we note only that $\tilde{p}(\cdot)$ is discontinuous on N so that the system is a generalized dynamical one.

The method outlined above can be used to treat system uncertainty as well as input disturbance. Generalizations to time and state-dependent matrices as well as to nonlinear systems may be found in Refs. 39-42.

We close the discussion with a simple example treated in Ref. 41. Consider a two-story building during an earthquake. For motion in one direction only, the equations of the linear model are (see Figure 4.1)

$$m_1(\ddot{y}_1 + \ddot{\delta}) = - c_1 \dot{y}_1 - k_1 y_1 + c_2(\dot{y}_2 - \dot{y}_1) + k_2(y_2 - y_1) + m_1 u_1 \tag{4.3}$$

$$m_2(\ddot{y}_2 + \ddot{\delta}) = - c_2(\dot{y}_2 - \dot{y}_1) - k_2(y_2 - y_1) + m_2 u_2$$

where m_i = mass of floor i

y_i = relative displacement of floor i

δ = ground displacement

$m_i u_i$ = applied force on floor i

k_i, c_i = positive constants

Equations (4.3) can be put into form (4.1) with

$$u = \begin{bmatrix} u_1 \\ u_2 \end{bmatrix} \qquad v = \begin{bmatrix} -\ddot{\delta} \\ -\ddot{\delta} \end{bmatrix}$$

In Ref. 43, the problem above is treated as one of optimal stochastic control; this results in a linear feedback control. Here we consider an actual disturbance, $\ddot{\delta}$, namely, the Northwest component of the Taft earthquake of 1952. Utilizing the data from Ref. 43 and letting $Q = I$, we deduce the nonlinear stabilizing control (4.2). Figure 4.2 shows the first floor displacement history corresponding to the nonlinear control (4.2) and to the linear control of Ref. 43, respectively.

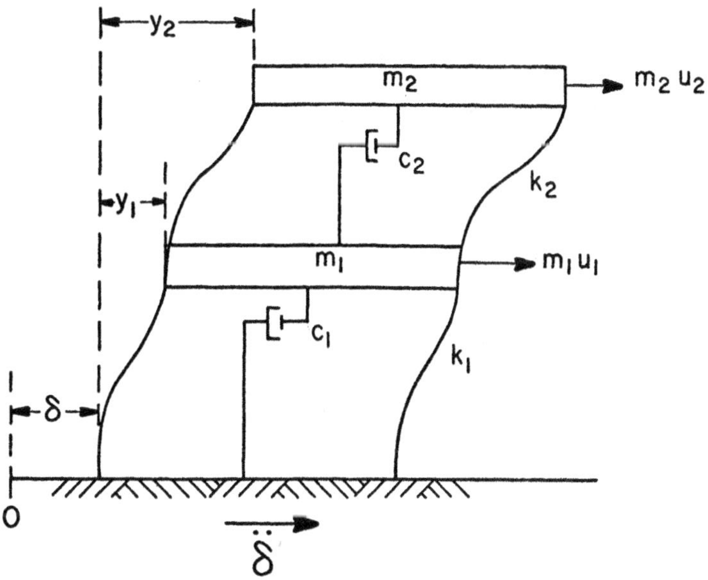

Figure 4.1, Two Story Building

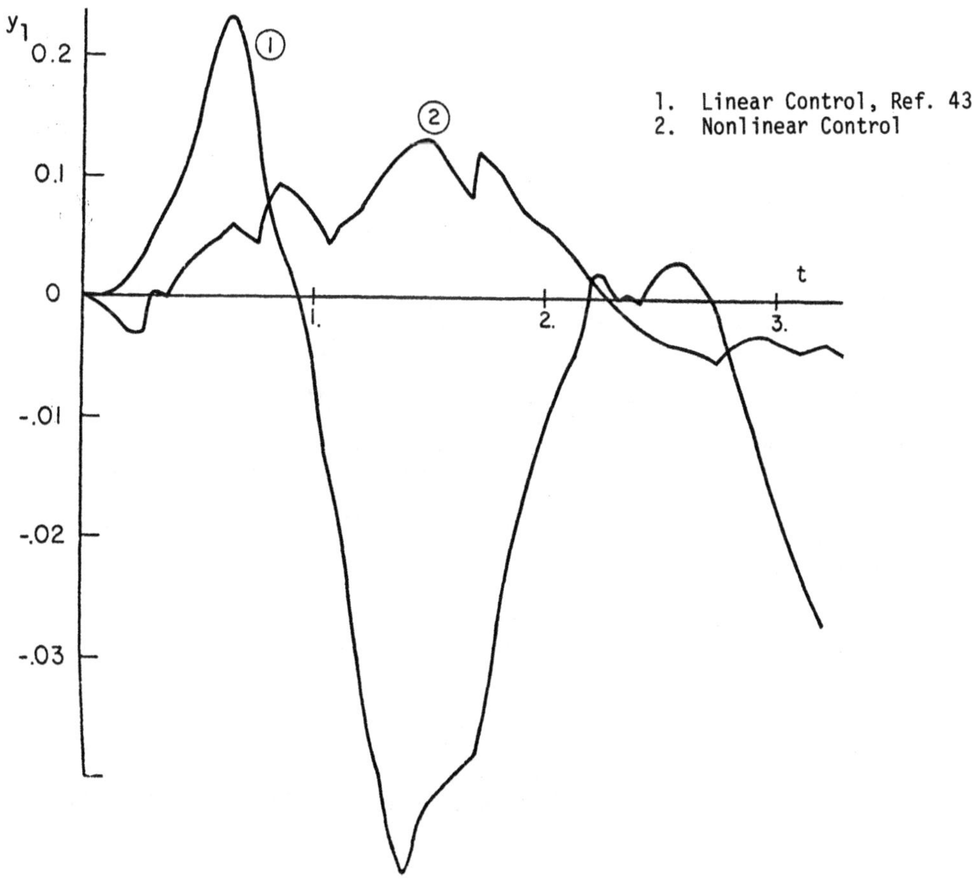

Figure 4.2, Displacement History of First Floor

References

1. Ho, Y. C., Differential Games, Dynamic Optimization, and Generalized Control Theory, J. of Optimization Theory and Appl., Vol. 6, No. 3, 1970.
2. Starr, A. W. and Ho, Y. C., Nonzero-Sum Differential Games, J. of Optimization Theory and Appl., Vol. 3, No. 3, 1969.
3. Starr, A. W. and Ho, Y. C., Further Properties of Nonzero-Sum Differential Games, J. of Optimization Theory and Appl., Vol. 3, No. 4, 1969.
4. Leitmann, G., Cooperative and Non-Cooperative Many Player Differential Games, CISM Monograph No. 190, Springer Verlag, Vienna, 1974.
5. Leitmann, G., (editor), Multicriteria Decision Making and Differential Games, Plenum Press, New York, 1976.
6. Leitmann, G. and Marzollo, A., (editors), Multicriteria Decision Making, CISM monograph No. 211, Springer Verlag, Vienna, 1975.
7. Nash, J., Non-Cooperative Games, Annals of Mathematics, Vol. 54, No. 2, 1951.
8. von Neumann, J., Zur Theorie der Gesellschaftsspiele, Mathematische Annalen, Vol. 100, pp. 295-320, 1928.
9. Karlin, S., Mathematical Methods and Theory in Games, Programming and Economics, Vol. 2, Infinite Games, Addison-Wesley, Reading, 1959.
10. Ponstein, J., An Extension of the Min-Max Theorem, SIAM Review, Vol. 7, No. 2, 1965.
11. Friedman, A., Differential Games, Wiley, New York, 1971.
12. Elliott, R. J. and Kalton, N. J., The Existence of Value in Differential Games, Memoir of American Math. Soc. 126, Providence, 1972.
13. Ho, Y. C. and Sun, F. K., Chapters XVI - XX in Multicriteria Decision Making and Differential Games, (edited by G. Leitmann), Plenum Press, New York, 1976.
14. Wilson, D. J., Mixed Strategy Solutions for Quadratic Games, J. of Optimization Theory and Appl., Vol. 13, No. 3, 1974.
15. Olsder, G. J., Information Structures in Differential Games, Proceed. of 2nd Kingston Conf. on Differential Games and Control Theory, M. Dekker, New York, 1976.
16. Filippov, A. F., Classical Solutions of Differential Equations with Multi-Valued Right-Hand Side, SIAM J. on Control, Vol. 5, No. 4, 1967.
17. Roxin, E., On Generalized Dynamical Systems Defined by a Contingent Equation, J. of Differential Equations, Vol. 1, pp. 188 f., 1965.
18. Isaacs, R., Differential Games, Wiley, New York, 1965.
19. Blaquière, A., Gérard, F. and Leitmann, G., Quantitative and Qualitative Games, Academic Press, New York, 1969.
20. Hájek, O., Pursuit Games, Academic Press, New York, 1975.
21. Yu, P. L., A Concept of Optimality in Differential Games, Int. J. of Systems Science, Vol. 7, No. 2, 1976.
22. Leitmann, G. and Rocklin, S., The Effect of Playability in Differential Games on the Relation Between Best Guaranteed Costs and Saddlepoint Values, J. of Optimization Theory and Appl., Vol. 15, No. 5, 1975.
23. Stalford, H., A Categorization of Two-Person Zero-Sum Differential Games, Naval Research Laboratory Report 7742, 1974.
24. Leitmann, G. and Mon, G., Some Geometric Aspects of Differential Games, J. of Astronautical Sciences, Vol. 14, No. 2, 1967.
25. Berkovitz, L. D., A Variational Approach to Differential Games, in Advances in Game Theory III, Princeton, 1964.
26. Berkovitz, L. D., Necessary Conditions for Optimal Strategies in a Class of Differential Games and Control Problems, SIAM J. Control, Vol. 5, No. 1, 1967.
27. Case, J. H., Toward a Theory of Many Player Differential Games, SIAM J. Control, Vol. 7, No. 2, 1969.
28. Leitmann, G. and Mon, G., On a Class of Differential Games, in Advanced Problems and Methods for Space Flight Optimization, (edited by Fraeijs de Veubeke), Pergamon Press, Oxford, 1969.
29. Blaquière, A. and Leitmann, G., Jeux Quantitatifs, Gauthiers-Villars, Paris, 1969.

30. Yu, P. L., Some Fundamental Qualifications of Optimal Strategies and Transition Surfaces in Differential Games, J. of Optimization Theory and Appl., Vol. 9, No. 6, 1972.
31. Pontryagin, L. S. et al., The Mathematical Theory of Optimal Processes, Interscience Publishers, New York, 1962.
32. Stalford, H. and Leitmann, G., Sufficiency Conditions for Nash Equilibria in N-Person Differential Games, in Topics in Differential Games, (edited by A. Blaquière), North-Holland, Amsterdam, 1973.
33. Clemhout, S. and Wan, H. Y., Jr., A Class of Trilinear Differential Games, J. of Optimization Theory and Appl., Vol. 14, No. 4, 1974.
34. Leitmann, G. and Schmitendorf, W. E., Profit Maximization Through Advertising: A Nonzero Sum Differential Games Approach, to appear.
35. Clemhout, S., Leitmann, G. and Wan, H. Y., Jr., A Differential Game Model of Oligopoly, J. of Cybernetics, Vol. 3, No. 1, 1973.
36. Leitmann, G. and Liu, P. T., A Differential Game Model of Labor Management Negotiations During Strike, J. of Optimization Theory and Appl., Vol. 13, No. 4, 1974 and Vol. 14, No. 4, 1974.
37. Clemhout, S., Leitmann, G. and Wan, H. Y., Jr., Bargaining Under Strike: A Differential Game View, J. of Economic Theory, Vol. 11, No. 1, 1975.
38. Clemhout, S., Leitmann, G. and Wan, H. Y., Jr., Equilibrium Patterns for Bargaining Under Strike: A Differential Game Model, in Directions in Decentralized Control, Many-Person Optimization, and Large Scale Systems, (edited by Y. C. Ho), Plenum Press, New York, 1976.
39. Gutman, S. and Leitmann, G., Stabilizing Control for Linear Systems with Bounded Parameter and Input Uncertainty, Proceed. of 7th IFIP Conf. on Optimization Techniques, Springer Verlag, Berlin, 1976.
40. Leitmann, G., Stabilization of Dynamical Systems Under Bounded Input Disturbance and Parameter Uncertainty, Proceed. of 2nd Kingston Conf. on Differential Games and Control Theory, M. Dekker, New York, 1976.
41. Gutman, S. and Leitmann, G., Stabilizing Feedback Control for Dynamical Systems with Bounded Uncertainty, Proceed. of IEEE Conf. on Decision and Control, Miami, 1976.
42. Gutman, S., Uncertain Dynamical Systems - A Differential Game Approach, NASA Technical Memorandum X-73, 135, 1976.
43. Yang, J. N., Application of Optimal Control Theory to Civil Engineering Structures, J. of Engineering Mechanics Division, ASCE, Vol. EM-6, December 1975.

On Observation Costs and Information Structures in Stochastic Differential Games.

Geert Jan Olsder
Department of Applied Mathematics
Twente University of Technology
Enschede, the Netherlands

1. Introduction.

In practical control problems as well as differential games difficulties are often encountered in obtaining information about the state of the dynamic system studied. Three principal difficulties are contamination by noise, time lag and a high cost of obtaining data. In this paper attention is focussed on the high costs of the data. The decisionmaker(s) decide(s) not only what plant control to employ, but also whether or not any particular observation is economically justified.
A number of authors, [1] - [4], have considered optimal control problems with observation costs. The current paper involves an extension to differential games. A two-person zero-sum game will be considered. Results are obtained for discrete time linear systems. At every stage careful attention must be given to the order in which the events occur. Crucial is also what one player gets to know about his opponent; does he for instance know whether his opponent made an observation ? Different results are obtained when the order of steps is changed and when the information about what the other player has done is altered.

In linear-quadratic optimal control problems it has been proved [4] that the optimal observation decisions are independent of the numerical values of the measurements and hence can be calculated off-line. It is conjectured that this is also true for linear-quadratic differential games; the proof has only been given for one-step discrete games.

Three different problems are considered (different in the sense of information known to the other player and the order of events):

(i) Both players decide at the beginning independently as to at which stages they will make an observation and then announce their decisions regarding the observation-policy to each other. Then the actual game is played and both players, of course, stick to the policy they have chosen (section 3).

(ii) At each stage the players first decide whether they will make an observation of the current state and subsequently, they choose the controls. Both these decisions are not announced to the other player (section 4).

(iii) At each stage the players decide independently whether they will make an observation of the current state and announce the decision to the other player. Subsequently the players choose their controls, which are not announced (section 5).

The information known to one player (via his measurements) is in general not known to the other player (who may have made measurements at different stages) and one therefore speaks of a non-nested information structure [5].

2. Problem statement.

We are given the system

$$x(k+1) = A(k)\ x(k) + B_p(k)\ u_p(k) + B_e(k)\ u_e(k) + F(k)\ \xi(k),\ k=0,1,\ldots,K-1. \tag{2.1}$$

The initial condition is $x(0) \sim N(x_o, P_o)$. The vectors x, u_p, u_e, ξ have n, r_p, r_e, q components respectively and $\xi(k)$ are independent Gaussian noises with

$$E\{\xi(k)\} = 0,\ E\{\xi(k)\ \xi'(l)\} = I\ \delta_{kl},\ I \text{ identity matrix,}$$

$$E\{x(0)\ \xi'(k)\} = 0,\quad k,\ l = 0,\ 1\ ,\ldots,\ K-1.$$

The matrices A, B_p, B_e F have appropriate dimensions.
The controls u_p and u_e will be chosen by the players P (of Pursuer) and E (of Evader) respectively. Player P has the possibility of observing

$$y_p(k) = C_p(k)\ x(k) + G_p(k)\ \xi(k), \tag{2.2}$$

and E can observe

$$y_e(k) = C_e(k)\ x(k) + G_e(k)\ \xi(k), \tag{2.3}$$

where the vectors y_p and y_e have m_p , m_e components respectively. The matrices C_p, C_e, G_p and G_e again have appropriate dimensions. Whether or not P makes an observation is indicated by the index $i_p(k)$. If P chooses $i_p(k) = 0$, then the observation $y_p(k)$ does not become known to P; if P chooses $i_p(k) = 1$, then he will observe $y_p(k)$. The index $i_e(k)$ is defined similarly. If P decides to observe $y_p(k)$, it will cost him an amount of $w_p(k)$; similarly the costs to E for observing $y_e(k)$ are $w_e(k)$; $w_p(k)$, $w_e(k) \geq 0$.

The numerical value of an observation for a player will never be announced to the other player. The fact that an observation took place may or may not be known to the other player, which will of course give rise to different solutions. It is tacitly assumed that the system characteristics Σ,

$$\Sigma \triangleq \{A,\ B_p,\ B_e,\ C_p,\ C_e,\ P_o,\ x_o,\ G_p,\ G_e,\ F,\ D,\ R_p,\ R_e,\ w_p,\ w_e\},$$

are known to both players. (D, R_p, R_e will be defined in the next paragraph).

P wants to minimize and E wants to maximize $E\{J(x(0),\ 0)\}$, where

$$J(x(i),i) = \frac{1}{2} \sum_{j=i}^{K-1} [x'(j+1)\ D(j+1)\ x(j+1) + u_p'(j)\ R_p(j)\ u_p(j) - u_e'(j)\ R_e(j)\ U_e(j) + + i_p(j)\ w_p(j) - i_e(j)\ w_e(j)],$$

and where it is assumed that $R_p(j)$, $R_e(j) > 0$. In order to state the problems in the forthcoming sections, the following definitions are made:

$$I_p(j) \triangleq \{i_p(0), \ldots, i_p(j)\} \quad ; \quad I_e(j) \triangleq \{i_e(0), \ldots, i_e(j)\} \quad ;$$

$$U_p(j) \triangleq \{u_p(0), \ldots, u_p(j)\} \quad ; \quad U_e(j) \triangleq \{u_e(0), \ldots, u_e(j)\} \quad ;$$

$$z_p(j) \triangleq i_p(j)\ y_p(j) \quad ; \quad z_e(j) \triangleq i_e(j)\ y_e(j) \quad ;$$

$$Z_p(j) \triangleq \{z_p(0), \ldots, z_p(j)\} \quad ; \quad Z_e(j) \triangleq \{z_e(0), \ldots, z_e(j)\} \quad .$$

In the information structure in section 3, P and E choose $I_p(K-1)$ and $I_e(K-1)$ respectively, independently of each other. Subsequently the game is played. The information available to P for a decision about $u_p(j)$ is $I_p(K-1)$, $I_e(K-1)$, $U_p(j-1)$, $Z_p(j)$ (and Σ). Similarly the information available to E in order to decide about $u_e(j)$ is $I_p(K-1)$, $I_e(K-1)$, $U_e(j-1)$ and $Z_e(j)$.

In the information structure in section 4, $i_p(j)$ may depend on $I_p(j-1)$, $U_p(j-1)$ and $Z_p(j-1)$, whereas $u_p(j)$ may depend on $I_p(j)$, $U_p(j-1)$ and $Z_p(j)$. Similarly $i_e(j)$ may depend on $I_e(j-1)$, $U_e(j-1)$ and $Z_e(j-1)$, and $u_e(j)$ may depend on $I_e(j-1)$, $U_e(j-1)$ and $Z_e(j)$. Given this information structure the players do not know whether their opponent has made a measurement.

In the information structure of section 5 the players do know at each stage whether their opponent has made an observation; $i_p(j)$ may depend on $I_p(j-1)$, $I_e(j-1)$, $U_p(j-1)$ and $Z_p(j-1)$; $u_p(j)$ may depend on $I_p(j)$, $I_e(j)$, $U_p(j-1)$ and $Z_p(j)$. Analogously $i_e(j)$ may depend on $I_p(j-1)$, $I_e(j-1)$, $U_e(j-1)$, $Z_e(j-1)$; $u_e(j)$ may depend on $I_p(j)$, $I_e(j)$, $U_e(j-1)$ and $Z_e(j)$.

The dependence relations as introduced above, must belong to a well defined class of functions, such as for instance the class of linear functions.

Remark. Instead of the game in section 3, the following nonzero-sum game is considered. The model, the measurements and the information structure remain the same. Now, however, P wants to minimize $E\{J_p\}$ and E wants to maximize $E\{J_e\}$, where

$$J_p \triangleq \frac{1}{2} \sum_{j=0}^{K-1} [x'(j+1)D(j+1)x(j+1) + u_p'(j)R_p(j)u_p(j) - u_e'(j)R_e(j)u_e(j) + i_p(j)w_p(j)];$$

$$J_e \triangleq \frac{1}{2} \sum_{j=0}^{K-1} [x'(j+1)D(j+1)x(j+1) + u_p'(j)R_p(j)u_p(j) - u_e'(j)R_e(j)u_e(j) - i_e(j)w_e(j)],$$

that is, the costs of observation are not to the advantage of the opponent. The Nash-solution to this nonzero-sum game equals the saddlepoint-solution of the corresponding zero-sum game of section 3. The argument is that $\Sigma i_e(j)\, w_e(j)$ is redundant in J_p, i.e. it is not influenced by any decision of P. Similarly $\Sigma i_p(j)\, w_p(j)$ is redundant in J_e. Obviously this argument will fail for the information structures of sections 4 and 5. □

3. The observation indices are chosen beforehand.

The sets $I_p(K-1)$ and $I_e(K-1)$ are chosen by P and E respectively, at the beginning of the game and independently of each other. Hence the game is open-loop w.r.t. the observation-indices. Subsequently the controls u_p and u_e are chosen. At stage j P chooses $u_p(j)$ based on the information available, $I_p(K-1)$, $I_e(K-1)$, $U_p(j-1)$, $Z_p(j)$. Similarly E chooses $u_e(j)$. Hence the game is closed loop w.r.t. the controls u_p and u_e.

Only a finite number of different sets $I_p(K-1)$ and $I_e(K-1)$ exist; that is 2^K for each. For each combination of a set $I_p(K-1)$ and a set $I_e(K-1)$ the quantity $\max_{u_e} \min_{u_p} E\,\{J(x(0),0)\}$ is calculated. In this way a matrix of size $2^K \times 2^K$ can be constructed. Once this has been done it is easy to calculate the optimal $I_p(K-1)$ and $I_e(K-1)$, that is by means of the well known theory of matrix games [7]. The solution, which in general will be mixed, is not necessarily unique, although the outcome of the game is unique.

In the remainder of this section we will concentrate on how to calculate $\max_{u_e} \min_{u_p} E\{J(x(0),\, 0)\}$, once a particular $I_p(K-1)$ and $I_e(K-1)$ are given. For the sake of simplicity we will assume $C_p(k) = C_e(k) = I$, $G_p(k) = G_e(k) = 0$, $k = 0, 1, \ldots, K-1$. This assumption reduces the number of calculations considerably. Another basic assumption is that $u_p(k)$ and $u_e(k)$ must be linear functions of $E\{x(0)\}$ and the observations, i.e.

$$u_p(k) = \sum_{j=0}^{k} N_p(k,j)\, i_p(j)\, x(j) + a_p(k)\, x_o ,$$

$$u_e(k) = \sum_{j=0}^{k} N_e(k,j)\, i_e(j)\, x(j) + a_e(k)\, x_o .$$

Implicit equations will be derived for the matrices $N_p(k,j)$, $N_e(k,j)$, $a_p(k)$ and $a_e(k)$. Those $N_p(k,j)$ and $N_e(k,j)$ for which the corresponding $i_p(j)$ resp. $i_e(j)$ is zero do not play a role and are assumed to be zero. Another reduction can be obtained in the number of the unknowns N_p, N_e, a_p and a_e to be determined.

<u>Theorem</u> Suppose $j \geq l$, $i_p(l) = i_p(l) = 1$ and $i_p(k).i_e(k) = 0$ for $k = l+1, l+2, \ldots, j$. Then

$$N_p(j,k) = 0,\; N_e(j,k) = 0, \quad k = 0, 1, \ldots, l-1,$$

$$a_p(j) = 0,\; a_e(j) = 0,$$

and in general $N_p(j,k) \neq 0$, $N_e(j,k) \neq 0$ f_n $k = l, l+1, \ldots, j$.

Proof. Because $i_p(l) = i_e(l) = 1$, both players know $x(l)$ exactly and hence the Markov-property holds, i.e. for $j \geq l$ the state $x(j)$ is completely determined by $u_p(k)$, $u_e(k)$, $\xi(k)$, $k \geq l$; the past is irrelevant. Suppose that for $k = l, l+1, \ldots, j$, the observation indices are as given in the figure:

$i_p(k)$	1	1	0	1	1	0	0	1	
	$j=l$	$l+1$	$l+2$	$l+3$	$l+4$	$l+5$	$l+6$	$j=l+7$	$\rightarrow k$
$i_e(k)$	1	0	1	0	0	1	0	0	

At stage $j(= l + 7)$ P knows $x(j)$ exactly; not so E, who will at least use his observation $x(l+5)$ in order to determine $u_e(j)$. Therefore P would like to know $x(l+5)$, or at least an estimate of $x(l+5)$. For that reason he needs $x(l+4)$ and $x(l+7)$. Player E knows therefore that $u_p(l+7)$ will not only depend on $x(l+7)$ but also on $x(l+4)$. E wants to make as good as possible an estimate of $x(i+4)$, for which he needs $x(l+2)$ and $x(l+5)$, and so on. In this way it is clear that in general $N_p(j,k) \neq 0$, $N_e(j,k) \neq 0$ for $k = l, l+1, \ldots, j$. □

In order to solve $N_p(k,j)$, $N_e(k,j)$, $a_p(k)$ and $a_e(k)$ two control problems will be considered. First it is assumed that $N_e(k,j)$ and $a_e(k)$ are fixed and then P faces an ordinary optimal control problem, the solution of which is expressed in terms of the unknown N_e, a_e. Thus N_p, a_p become functions of N_e, a_e. Reversing the roles of the players, i.e. now fixing N_p, a_p, E faces an optimal control problem as well and another set of equations is obtained; N_e, a_e as functions of N_p, a_p. Together these two sets of equations constitute implicit equations for N_p, N_e, a_p and a_e, which can be solved in principle.

Suppose that $N_e(k,j)$, $a_e(k)$ are fixed. The optimal control problem faced by P can be put into the standard linear-quadratic form in the following way. Define vectors $\pi(k)$, $k = 0, 1, \ldots, K$, each of $(K+1)n$ components as follows. The first n components equal vector $x(0)$, the second set of n components equals vector $x(1)$, and so on up to the $(k+1)$st set of n components which equals $x(k)$. The remaining sets of n components equal $x(k)$ as well. With respect to this new state vector, the evolution of the system can be written as linear, inhomogeneous equations:

$$\pi(k+1) = \Phi_p(k)\,\pi(k) + \Gamma_p(k)\,u_p(k) + Y_p(k)\,\xi(k) - \Gamma_e(k)\,a_e(k)\,x_o, \qquad (3.1)$$

$$k = 0, 1, \ldots, K-1.$$

The matrices introduced are defined as follows. $\Phi_p(k)$ is a matrix partitioned in blocks of $n \times n$ submatrices. The l,j-th block is denoted by $(\Phi_p(k))_{l,j}$. The matrix $Y_p(k)$ is also partitioned in blocks of n × n submatrices. The j-th block is denoted by $(Y_p(k))_j$. Similar partitioning and labeling of the submatrices with respect to the other matrices.

$$(\Phi_p(k))_{j,j} = I,\ j = 1, \ldots, k+1;$$

$$(\Phi_p(k))_{j,j} = A(k),\ j = k+2, \ldots, K+1;$$

$$(\Phi_p(k))_{1,j} = B_e(k)\ N_e(k, j-1)\ i_e(j-1), \qquad 1 \geq k+2 \text{ and } j \leq k+1;$$

$$(\Phi_p(k))_{1,j} = 0 \text{ elsewhere.}$$

$$(\Gamma_p(k))_j = 0;\ (\Gamma_e(k))_j = 0;\ (Y(k))_j = 0;\ i = 1, 2, \ldots, k+1;$$

$$(\Gamma_p(k))_j = B_p(k);\ (\Gamma_e(k))_j = B_e(k);\ (Y(k))_j = I;\ j = k+2, \ldots, K+1.$$

The initial condition becomes

$$E(\pi(0)) = (x_o \vdots x_o \vdots \ldots \vdots x_o)' \triangleq \pi_o';$$

$$E\{(\pi(0) - \pi_o)\ (\pi(0) - \pi_o)'\} \triangleq Q_o,$$

where Q_o is a blockmatrix of which all blocks are equal to P_o. With respect to the newly defined state variable the criterion - apart from the observation costs and other terms which are not u_p-dependent - becomes

$$J = \frac{1}{2} \sum_{j=0}^{K-1} [\pi'(j+1)\ \Delta(j+1)\ \pi(j+1) + u_p'(j)\ R_p(j) u_p(j) - 2\ x_o'\ a_e'(j+1)\ R_e(j+1)\ \Lambda_e'(j+1)\pi(j+1)],$$

where

$$(\Lambda_e(j))_1 = N_e(j, 1-1)\ i_e(1-1),\ 1 = 1, \ldots, j+1;$$

$$(\Lambda_e(j))_1 = 0 \qquad , 1 = j+2, \ldots, K+1;$$

$$\Delta(j) = \tilde{N}_e(j)\ R_e(j)\ \tilde{N}_e'(j) + \tilde{D}(j),\ j = 1, \ldots, K-1;$$

$$(\tilde{D}(j))_{k,1} = 0, \text{ except for } k = 1 = j+1; \qquad (\tilde{N}_e(j))_1 = N_e(j,1),\ 1 = 1, \ldots, j;$$

$$(\tilde{D}(j))_{j+1,\ j+1} = D(j); \qquad (\tilde{N}_e(j))_1 = 0, \text{ elsewhere.}$$

$$\Delta(K) = \tilde{D}(K);$$

Now a standard linear quadratic optimal control problem has been formulated. By means of the Kalman filtering theory the minimum variance estimate $\hat{\pi}(k)$ of $\pi(k)$ can be expressed in terms of the observations $y_p(k)$;

$$y_p(k) = \chi_p(k)\ \pi(k),$$

where

$$(\chi_p(k))_j = I\ i_p(j-1),\ j = k;$$

$$(\chi_p(k))_j = 0, \text{ elsewhere.}$$

These expressions are linear functions of the observations. Subsequently the theory of the optimal regulator problem is used to express $u_p(k)$ as a (linear) function of $\hat{\pi}(k)$. Thus $u_p(k)$ becomes a linear function of the observations and that is exactly the form we assumed. The general case, in which we do not have the restrictions $C_p(k) = C_e(k) = I$, $G_p(k) = G_e(k) = 0$, can be treated along the same lines. The theorem of this section is not valid any more for the general case. The state space description (3.1) has also to be adapted for that case; see for instance [6].

4. Players do not know whether their opponent has made an observation.

This problem will be illustrated by a simple one-stage scalar example:

$$x(1) = x(0) + u_p(0) + u_e(0) + \xi(0) \ ;$$
$$x(0) \sim N(x_o, \sigma^2) \ ; \qquad (4.1)$$
$$\xi(0) \sim N(0, 1) \ ;$$

$x(0)$ and $\xi(0)$ are uncorrelated. The criterion is

$$\min_{i_p(0),\, u_p(0)} \quad \max_{i_e(0),\, u_e(0)} \quad E\{J\}, \qquad (4.2)$$

where

$$J = x^2(1) + u_p^2(0) - c\, u_e^2(0) + i_p(0)\, w_p(0) - i_e(0)\, w_e(0) \ , \qquad (4.3)$$

c is a positive constant greater than 1. Each player first decides whether he will make an observation or not. Subsequently he chooses his control $u_p(0)$ resp. $u_e(0)$. If a player has made an observation, he may, of course, use the obtained measurement to improve the control. Each player does not get to know whether his opponent has made an observation. If a player makes an observation he gets to know $x(0)$ exactly. Suppose that the chance of P making an observation is r, $0 \le r \le 1$; the chance of not making an observation is $1 - r$. Similarly the chance of E (not) making an observation is supposed to be q $(1-q)$; $0 \le q \le 1$. If P has not made an observation, $u_p(0)$ is assumed to be of the form

$$u_p(0) = \alpha_1\, x_o + \alpha_2, \qquad (4.4)$$

If P has made an observation, then it is assumed that $u_p(0)$ has the form

$$u_p(0) = \alpha_3\, x_o + \alpha_4\, x(0) + \alpha_5, \qquad (4.5)$$

where $x(0)$ is the exact value of the initial state, obtained by the measurement. Similarly for E these forms are respectively

$$u_e(0) = \beta_1\, x_o + \beta_2, \qquad (4.6)$$

$$u_e(0) = \beta_3\, x_o + \beta_4\, x(0) + \beta_5 \qquad (4.7)$$

Suppose P has not made an observation. He knows that E will play (4.6) with chance 1-q and (4.7) with chance q. Player P does not know that value of $x(0)$ in (4.7). If (4.5) is substituted in $E\{J\}$, the result, to be denoted by $J_p(i_p, i_e) = J_p(0,0)$, is

$$J_p(0,0) = x_o^2 + \sigma^2+1+(\beta_1 x_o+\beta_2)^2(1- c+2x_o(\beta_1 x_o+\beta_2) + +2u_p^2(0)+2u_p(0)\ (x_o+\beta_1 x_o+\beta_2), \quad (4.8)$$

whereas, if (4.7) is substituted in E{J}, we get

$$J_p(0,1) = x_o^2+\sigma^2+1+(1-c)\{\beta_3^2 x_o^2+\beta_4^2(x_o^2+\sigma^2)+\beta_5^2+2\beta_3\beta_4 x_o^2+2\beta_3\beta_5 x_o+2\beta_4\beta_5 x_o\}+ + 2\beta_3 x_o^2+2\beta_4(x_o^2+\sigma^2)+2\beta_5 x_o-w_e(0)+ + 2u_p^2(0)+2u_p(0)\ (x_o+\beta_3 x_o+\beta_4 x_o+\beta_5). \quad (4.9)$$

Player P expects the outcome of the criterion to be

$$(1-q)\ J_p(0,0) + q\ J_p(0,1).$$

The minimum of this expression is achieved for

$$u_p(0) = -\frac{1}{2}[\{(1+\beta_1)(1-q)+(1+\beta_3+\beta_4)q\}x_o+\beta_2(1-q)+\beta_5 q]. \quad (4.10)$$

and will be denoted by $(1-q)\ J_p^*(0,0) + q\ J_p^*(0,1)$.

Now suppose that P has made an observation. If (4.6) and (4.7) are substituted in E{J}, we get respectively:

$$J_p(1,0) = x^2(0)+(\beta_1 x_o+\beta_2)^2(1-c)+2x(0)(\beta_1 x_o+\beta_2)+w_p(0)+1+ + 2u_p^2(0)+2u_p(0)(x(0)+\beta_1 x_o+\beta_2); \quad (4.11)$$

$$J_p(1,1) = x^2(0)+(\beta_3 x_o+\beta_4 x(0)+\beta_5)^2(1-c)+2x(0)(\beta_3 x_o+\beta_4 x(0)+\beta_5)+1+ + w_p(0)-w_e(0)+2u_p^2(0)+2u_p(0)(x(0)+\beta_3 x_o+\beta_4 x(0)+\beta_5). \quad (4.12)$$

Minimization of $(1-q)\ J_p(1,0) + q\ J_p(1,1)$ w.r.t. $u_p(0)$ yields

$$u_p(0) = \frac{1}{2}[\{(1-q)\beta_1+q\beta_3\}x_o+\{1+\beta_4 q\}x(0)+\{\beta_2(1-q)+\beta_5 q\}]; \quad (4.13)$$

the minimum value is denoted by $(1-q)\ J_p^*(1,0) + q\ J_p^*(1,1)$.

If instead of the viewpoint of P, we now consider the viewpoint of E, we find similar expressions for $J_e(i_p,\ i_e)$, $i_p,\ i_e = 0,1$. If player E has not made an observation he expects the outcome of the criterion to be $(1-r)\ J_e(0,0) + r\ J_e(1,0)$, which is minimized for

$$u_e(0) = -\frac{1}{1-c}[\{(1-r)(1+\alpha_1)+r(1+\alpha_3+\alpha_4)\}x_o+(1-r)\alpha_2+r\alpha_5]. \quad (4.14)$$

The minimum is denoted by $(1-r)\ J_e^*(0,0)+ rJ_e^*\ (1,0)$. Similarly, if E has made an observation, $(1-r)\ J_e(0,1) + r\ J_e(1,1)$ is minimized by

$$u_e(0) = -\frac{1}{1-c}[\{(1-r)\alpha_1+p\alpha_3\}x_o+\{1+r\alpha_4\}x(0)+\{(1-r)\alpha_2+r\alpha_5\}], \quad (4.15)$$

yielding as minimum $(1-r)\ J_e^*\ (0,1) + r\ J_e^*\ (1,1)$.

The unknowns α_i, β_i, $i = 1, 2,\ldots,5$, can be solved from (4.4) - (4.7), (4.10), (4.13) - (4.15) and expressed in r, q;

$$\begin{pmatrix} 2 & 0 & 0 & 1-q & q & q \\ 0 & 2 & 0 & 1-q & q & 0 \\ 0 & 0 & 2 & 0 & 0 & q \\ 1-r & r & r & 1-c & 0 & 0 \\ 1-r & r & 0 & 0 & 1-c & 0 \\ 0 & 0 & r & 0 & 0 & 1-c \end{pmatrix} \begin{pmatrix} \alpha_1 \\ \alpha_3 \\ \alpha_4 \\ \beta_1 \\ \beta_3 \\ \beta_4 \end{pmatrix} = - \begin{pmatrix} 1 \\ 0 \\ 1 \\ 1 \\ 0 \\ 1 \end{pmatrix}, \qquad (4.16)$$

which yields

$$\begin{aligned} \alpha_1 &= - c\beta_1 \quad ; \quad \beta_1 = - \frac{1}{1-2c} \quad ; \\ \alpha_3 &= \alpha_1 - \alpha_4 \quad ; \quad \beta_3 = \beta_1 - \beta_4 \quad ; \\ \alpha_4 &= \frac{1}{2}(- 1 - q\beta_4) \; ; \quad \beta_4 = \frac{r-2}{2(1-c)-rq} \quad ; \end{aligned} \qquad (4.17)$$

and the solution of the remaining unknowns is

$$\alpha_2 = \alpha_5 = \beta_2 = \beta_5 = 0.$$

Define

$$\begin{aligned} f_p(r,q) &= (1-r)(1-q)J_p^*(0,0)+(1-r)qJ_p^*(0,1)+r(1-q)J_p^*(1,0)+rq\, J_p^*(1,1) \; ; \\ f_e(r,q) &= (1-r)(1-q)J_e^*(0,0)+(1-r)qJ_e^*(0,1)+r(1-q)J_e^*(1,0)+rq\, J_e^*(1,1) \; , \end{aligned} \qquad (4.18)$$

in which (4.17) has been substituted. This yields

$$\begin{aligned} f_p(r,q) &= \sigma^2+1+ \frac{c}{2c-1}\, x_o^2 +q[(1-c)\beta_4^2\sigma^2+2\beta_4\sigma^2-w_e(0)]+r[- \tfrac{1}{2}(1+\beta_4 q)^2\sigma^2+w_p(0)] \; , \\ f_e(r,q) &= \sigma^2+1+ \frac{c}{2c-1}\, x_o^2+r[2\alpha_4^2\sigma^2+2\alpha_4\sigma^2+w_p(0)]+q[- \tfrac{1}{1-c}(1+\alpha_4 r)^2\sigma^2-w_e(0)] \quad . \end{aligned} \qquad (4.19)$$

Because of the zero-sum character, we should have $f_p \equiv f_e$, which turns out to be true. P wants to minimize $f_p(r,q)$ w.r.t. r, subject to $0 \le r \le 1$ and E wants to maximize $f_p(r,q)$ w.r.t. q, subject to $0 \le q \le 1$. Note that the optimal r and q are independent of x_o! It is easily shown that $\partial^2 f_p/\partial r^2 > 0$ and $\partial^2 f_p/\partial q^2 < 0$ on the square $0 \le r, q \le 1$. Hence the game with the (continuous) payoff $f_p(r,q)$ on the square $0 \le r, q \le 1$ is a convex-concave game and for such a game it is known that they have pure optimal strategies [7].

If the optimal strategiesr,q are inside the square $[0,1] \times [0,1]$, then they satisfy $\partial f_p/\partial r = 0$, $\partial f_p/\partial q = 0$, which yields

$$r = \sqrt{2}\; \frac{\sigma - \sqrt{(c-1)w_e(0)}}{\sqrt{w_p(0)}} \quad ; \quad q = \frac{\sqrt{2(c-1)w_p(0)}-\sigma}{\sqrt{w_e(0)}}$$

Remark. Eqs. (4.10), (4.13) - (4.15) form the solution to the following problem: r and q are given numbers, $0 \le r, q \le 1$; $1 - r(1-q)$ is the chance that at any particular stage the true state of the system will not be known to player P(E). This is the differential game extension of a control problem considered in [8]. □

Remark. The extension to K-stages becomes extremely complicated, though the principle of solution remains the same. To show some of the complications, the case K = 2 will be considered in some detail. The scalar system is given by

$$x(j+1) = x(j) + u_p(j) + u_e(j) + \xi(j),\ j = 0,\ 1;$$

$$x(0) \sim N(x_o,\ \sigma^2),$$

$$\xi(j) \sim N(0,\ 1),\ j = 1,\ 2;$$

$x(0)$, $\xi(0)$, $\xi(1)$ are mutually uncorrelated. Player P wants to minimize $E\{J\}$, player E wants to maximize $E\{J\}$, J given by

$$J = x^2(2) + x^2(1) + u_p^2(0) + u_p^2(1) - c\ u_e^2(0) - c\ u_e^2(1)$$

$$+ i_p(0)\ w_p(0) + i_p(1)\ w_p(1) - i_e(0)\ w_e(0) - i_e(1)\ w_e(1),$$

where c is a positive constant. At each stage each player first decides whether he will make an (exact) observation (of the state) or not, and subsequently each player chooses his control. Suppose that at stage j the chance of P making an observation is $r(j)$, $j = 0,\ 1$; for E this is $q(j)$. Suppose for P that $i_p(0) = 0$ or 1 respectively, then $u_p(0)$ is assumed to be of the form

$$u_p(0) = \alpha_1\ x_o\ ,$$

$$u_p(0) = \alpha_2\ x_o + \alpha_3\ x(0)\ ,$$

respectively. If $i_p(0) = 0$, then at stage 1, either $i_p(1) = 0$ or $i_p(1) = 1$, which will give rise to $u_p(1)$ of the form

$$u_p(1) = \alpha_4\ x_o$$

$$u_p(1) = \alpha_5\ x_o + \alpha_6\ x(1)$$

respectively. If $i_p(0) = 1$, then also either $i_p(1) = 0$ or $i_p(1) = 1$, which yields the forms

$$u_p(1) = \alpha_7\ x_o + \alpha_8\ x(0)$$

$$u_p(1) = \alpha_9\ x_o + \alpha_{10}\ x(0) + \alpha_{11}\ x(1)$$

respectively. Similarly parameters $\beta_1 - \beta_{11}$ are introduced if one starts from E's viewpoint. In order to solve for $\alpha_1 - \alpha_{11}$, $\beta_1 - \beta_{11}$ one has to solve a system of 22 equations with 22 unknowns (which will now be nonlinear) (for K = 1 the corresponding system was (4.16), for general K the number of unknowns is $2((K+1)2^K-1))$. The optimal $r(1)$, $q(1)$ can be expressed as functions of $r(0)$, $q(0)$. Once this has been done one constructs a function $f(x_o,\ r(0),\ q(0))$, the analogue of (4.18). This function must be minimized by $r(0)$ and maximized by $q(0)$, subject to $0 \le r(0), q(0) \le 1$, which yields the optimal $r(0)$, $q(0)$. It is conjectured that the optimal $r(j)$ and $q(j)$, $j = 0\ ,\ldots,\ K-1$ do not depend on x_o. □

5. Players do know whether the opponent has made an observation.

This information structure will be elucidated by means of the example of the previous section. First both players decide independently whether to make an observation or not and then this fact is announced to the other player. If, for instance, the decisions are $i_p(0) = i_e(0) = 0$, then subsequently the criterion is minimized by $u_p(0)$ and maximized by $u_e(0)$. The optimal value of the criterion is then denoted by a_{11}. In the same way a_{12} is defined; this is the optimal value of the criterion if $i_p(0) = 0$, $i_e(0) = 1$; a_{21} corresponds to $i_p(0) = 1$, $i_e(0) = 0$ and a_{22} corresponds to $i_p(0) = 1$, $i_e(0) = 1$. Once a_{kj} are known, the following matrix game must be solved w.r.t. $i_p(0)$, $i_e(0)$:

$i_e(0)$ \ $i_p(0)$	0	1
0	a_{11}	a_{12}
1	a_{21}	a_{22}

We will concentrate on how to find the entries a_{ij}.

(i) $i_p(0) = i_e(0) = 0$. The criterion can be written as

$$\min_{u_p(0)} \max_{u_e(0)} [x_o^2+\sigma^2+2x_o(u_p(0)+u_e(0))+1+u_p^2(0)-cu_e^2(0)+(u_p(0)+u_e(0))^2],$$

which yields $u_p(0) = x_o\, c/(1-2c)$; $u_e(0) = -x_o/(1-2c)$, which in its turn yields $a_{11} = \sigma^2 + 1 - x_o^2\, c/(1-2c)$.

(ii) $i_p(0) = i_e(0) = 1$. The criterion can be written as

$$\min_{u_p(0)} \max_{u_e(0)} [x(0)+u_p(0)+u_e(0))^2+1+u_p^2(0)- u_e^2(0)+w_p(0)-w_e(0)],$$

which yields $u_p(0) = x(0)\, c/(1-2c)$; $u_e(0) = -x(0)/(1-2c)$ and

$$a_{22} = -(x_o^2 + \sigma^2)\, c/(1-2c) + 1 + w_p(0) - w_e(0).$$

(iii) $i_p(0) = 1$, $i_e(0) = 0$. Because E does not get any additional information about $x(0)$, it is easily seen that E will play $u_e(0) = -x_o/(1-2c)$, which is the same strategy as in the case $i_p(0) = i_e(0) = 0$. Of course P will realize this strategy of E and hence P will want to minimize

$$(x(0)+u_p(0) - \frac{x_o}{1-2c})^2+1+u_p^2(0) - \frac{cx_o}{(1-2c)^2} + w_p(0),$$

which yields $u_p(0) = (x_o/(1-2c) - x(0))/2$. The value of the criterion becomes

$1 + \frac{1}{2}\sigma^2 + w_p(0) - x_o^2 c/(1-2c)$.

(iv) $i_p(0) = 0$, $i_e(0) = 1$. Because in this case P does not get any additional information, he will play $u_p(0) = x_o\, c/(1-2c)$ and accordingly E will play in such a way as to maximize

$$(x(0) + \frac{x_o c}{1-2c} + u_e(0))^2 + 1 + (\frac{x_o c}{1-2c})^2 - c\, u_e^2(0) - w_e(0),$$

which yields $u_e(0) = (x(0) + x_o c/(1-2c))/(c-1)$ and hence

$a_{12} = \sigma^2 c/(c-1) + 1 - w_e(0) - x_o^2\, c/(1-2c)$.

If the term $- 1 + x_o^2\, c/(1-2c)$ is added to the components of the matrix $\{a_{kj}\}$ - this will not change the matrix game - the following matrix game results:

$i_e(0)$ \ $i_p(0)$	0	1
0	σ^2	$\frac{1}{2}\sigma^2 + w_p(0)$
1	$\sigma^2 \frac{c}{c-1} - w_e(0)$	$-\sigma^2 \frac{c}{1-2c} + w_p(0) - w_e(0)$

Note that also in this case the decision concerning whether to make an observation does not depend on x_o. Mixed strategies only occur if $w_p(0)$ and $w_e(0)$ simultaneously satisfy

$$\frac{1}{2} < \frac{w_p(0)}{\sigma^2} < \frac{c^2}{(c-1)\ (2c-1)} \tag{5.1}$$

$$\frac{1}{2(2c-1)} < \frac{w_e(0)}{\sigma^2} < \frac{1}{c-1} \tag{5.2}$$

If (5.1) and (5.2) are satisfied, then

$$r = \frac{a_{21} - a_{11}}{-a_{11}+a_{12}+a_{21}-a_{22}}; \qquad q = \frac{a_{12} - a_{11}}{-a_{11}+a_{12}+a_{21}-a_{22}}.$$

<u>Numerical example</u>. Suppose $\sigma^2 = 1$, $c = 2$, $w_p(0) = w_e(0) = 3/4$. The optimal r, q are $r = \frac{3}{10}$, $q = \frac{3}{10}$. With the same system characteristics and the information structure of section 4 one gets as optimal values $r = .2188$, $q = .2595$. □

If another example is chosen such that at least one of the four inequalities of (5.1) and/or (5.2) is not satisfied, then the optimal solution will be $r = 0$ or 1, $q = 0$ or 1. W.r.t. the information structure (i.s.) of section 4 it is also possible that solutions exist such as for instance $r = 1$, $0 < q < 1$. W.r.t. the i.s. of this section the optimal solution is situated either inside the square $0 \le r, q \le 1$ or at a corner.

In an example in which both (5.1) and (5.2) are violated, the solution w.r.t. the i.s. of this section equals the solution w.r.t. the i.s. of section 4. If, for instance, $w_p/\sigma^2 < \frac{1}{2}$ and $w_e/\sigma^2 > 1/(c-1)$, it is easily shown that the optimal solution w.r.t. the i.s. of this section is $r = 1$, $q = 0$. It follows from (4.19) that, subject to the same conditions, $\frac{\partial f_p(r,0)}{\partial q} < 0$ for $0 \le r \le 1$ and $\frac{\partial f_p(1,q)}{\partial r} < 0$ for $0 \le q \le 1$. This together with the convexity of f_p w.r.t. r and concavity w.r.t. q makes that $\min_r \max_q f_p(r, q)$ is achieved for $r = 1$, $q = 0$.

The K-stage game can be solved along the same lines as the K-stage game of the previous section. The complexity seems of the same kind.

6. A game easy to play.

A game which can easily be played by two players and a third person, acting both as a random generator and the messenger of information is described in this section. It shows the complexities of the games considered in this paper.

The discrete time game is played on four positions on a circle, see figure. During each stage each player can move a given point mass 90°, 0° or -90°. So if both players choose 90°, the point mass moves to the diametrically opposite point. Apart from the players' actions there is at each stage a stochastic disturbance (to be played by the third person), which is 0° with probability $\frac{1}{2}$, $\pm 90^\circ$ with probability $\frac{1}{4}$. The initial point is distributed, too, at position 2 (see the figure) with probability $\frac{1}{2}$ and either at position 1 or 3 with probabilities $\frac{1}{4}$. The number of stages is K. Let V denote the (absolute) distance measured in degrees from the point's position at the last stage from position one (hence this distance is maximally 180). Then P wants to minimize and E wants to maximize.

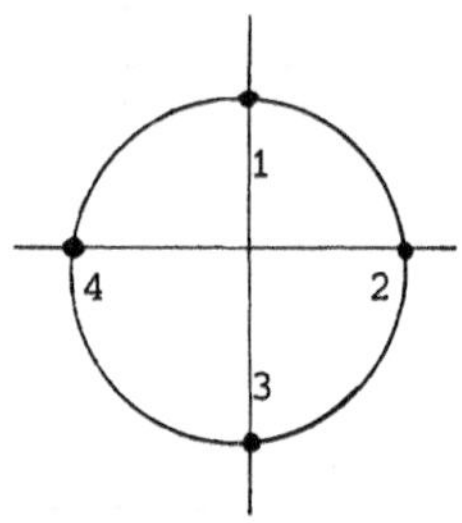

$$V + \Sigma\, i_p(k)\, w_p(k) - \Sigma\, i_e(k)\, w_e(k)$$

The rules of each of the sections 3, 4, and 5 can be employed. One can easily show that in general both the observation indices and the controls will become mixed strategies. One can formulate other games with more positions on the circle.

Acknowledgement. I am indebted to Professor Richard Longman of Columbia University, N.Y.,who drew my attention to control problems with observation costs.

References

[1] L. Meyer III, J. Peschon R.M. Dressler, "Optimal Control of Measurements Sub-

systems"m IEEE Transactions on Aut. Control, vol. AC-12, nr.5, 1967, pp 528-536.

[2] M. Aoki, M.T. Li, "Optimal Discrete-Time Control System with Cost for Observation", IEEE Transact. on Aut. Control, vol AC-14, nr. 2, 1969,pp 165-175.

[3] C.A. Cooper, N.E. Nahi, "An Optimal Stochastic Control Problem with Observation Cost", IEEE Transact on Aut. Control, vol. AC-16, 1971, pp 185-189.

[4] N.M. Olgac, C.A. Cooper, R. Longman, "Time Lagged Stochastic Control Problems with Observation Cost", Proceedings of the JACC 1976, West Lafayette, Indiana.

[5] Y.C. Ho, K.C. Chu, "Information Structure in Dynamic Multi-Person Control Problems", Automatica, vol. 10, nr. 4, 1974, pp. 341-351.

[6] W.W. Willman, "Formal solutions for a Class of Stochastic Pursuit-Evasion Games", IEEE Transactions on Autom. Control, AC-14, nr. 5, 1969, pp. 504-509.

[7] G. Owen, Game Theory, Saunders, 1968.

[8] R. Bellman, R. Kalaba, "A Note on Interrupted Stochastic Control Processes", Information and Control, no. 4, 1969, pp. 346-349.

DIFFERENTIAL GAMES WITH PARTIAL DIFFERENTIAL EQUATIONS

Emilio O. Roxin

University of Rhode Island
Department of Mathematics
Kingston, R.I. 02881, USA

SUMMARY: Examples are given to show that differential games involving partial differential equations can be of interest to applications. An example is given in which neither independent variable has the meaning of "time", showing that more general information structures are possible and reasonable. Discretization of the partial differential equation can lead from the optimality condition for ordinary differential games to a similar principle for partial differential games. Another such possibility is to use Fourier expansion. In the case of the heat and the wave equation, direct solution of some games show particular features of interest, similar to the ones known for ordinary differential games.

1. EXAMPLES OF APPLIED PROBLEMS

Optimal control problems involving partial differential equations (so called "distributed parameter systems") have been considered since the 1960's and the basic theory is already well established (see, for example, Butkovskiy [1], Komkov [8], Lions [9]). They have been found of importance to practical applications. That differential games involving partial differential equations ("partial differential games") can also be of practical interest will be shown by the following two examples.

A. POLLUTION CONTROL. Consider a lake, which for simplicity will be described as one-dimensional: $0 \leq x \leq 1$. Assume that an industry ("polluter") wishes to discharge a certain amount of pollutant into that lake in the time interval $[0,T]$. Assume that this pollutant diffuses and is decomposed naturally in such a way that its concentration $w(x,t)$ satisfies the equation

$$w_t(x,t) = a\, w_{xx}(x,t) - k\, w(x,t) + u(x,t).$$

Here u(x,t) is the rate of discharge of pollutant, and the total discharge may be normalized as

$$\int_0^1 \int_0^T u(x,t)\, dt\, dx = 1.$$

In order to discourage pollution, the Government measures w(x,t) at N samples (ξ_i,τ_i) within $[0,1]\times[0,T]$, and charges the industry a penalty

$$J = C \sum_{i=1}^{N} w(\xi_i,\tau_i), \quad \text{or} \quad J = C \max_i w(\xi_i,\tau_i).$$

With initial conditions w(x,0) = 0 and boundary conditions $w_x(0,t) = w_x(1,t) = 0$, and assuming N fixed, the problem arises to determine the best locations (ξ_i,τ_i) in order to maximize J, and the most convenient rate of discharge u(x,t) in order to minimize J.

For this problem to be completely defined, the information structure should be given. If, for example, the Government must announce in advance the locations (ξ_i,τ_i), then the problem is really a min-max and not a differential game in the usual sense of continuous update of the information. On the other hand, if each player keeps his actions secret, then it is reasonable to assume that the Government only knows the past measurements $w(\xi_i,\tau_i)$ but does not know $u(x,\tau)$ for all $\tau<t$. In such a case the game is with imperfect information.

Generalizations of this game to two space-dimensions, variable coefficients a(x), k(x) and continuous measurements $w(\xi(t),t)$ are easy to imagine.

B. COMPETITIVE FISHING. Consider a closed body of water (lake, sea) as a region D in the x^1-x^2 plane, and w(x,t) the "density" of fish in it. Let two contries (or two competing parties) fish in D and designate by u(x,t) and v(x,t) their fishing efforts. The amount of fish caught by "player u" in the time interval [0,T] may be given by an expression of the form

$$\int_0^T \int_D \phi(x)\, w(x,t)\, u(x,t)\, dx\, dt,$$

where $\phi(x)$ is an efficiency factor depending on location. (If player "u" is not allowed to fish in some parts of D, $\phi(x)$ should be set =0 there.) A similar expression should be valid for the catch of player "v".

If k is a multiplication factor by natural reproduction of fish, and their movement is taken into account as a kind of diffusion, the equation describing the time-evolution of w(x,t) will be of the following type

$$w_t(x,t) = a\,\Delta w(x,t) + [k - \phi(x)\,u(x,t) - \psi(x)\,v(x,t)]w(x,t),$$

with suitable initial and boundary conditions.

Some constraints will be imposed on u(x,t) and v(x,t) (first at all $u,v \geq 0$), and the coefficients a, k may be taken as depending on location and season.

The reward for each player will be the catch minus the total fishing effort, hence

$$J_1 = \iint \phi(x)\,w(x,t)\,u(x,t)\,dx\,dt - \iint u(x,t)\,dx\,dt$$

$$J_2 = \iint \psi(x)\,w(x,t)\,v(x,t)\,dx\,dt - \iint v(x,t)\,dx\,dt$$

This game is therefore non-zero-sum. Its solution will involve finding possible Nash-equilibria as well as Pareto-optimal points.

2. PARTIAL DIFFERENTIAL GAMES WITH GENERAL INFORMATION PATTERN

Differential games involving partial differential equations generalize the ones with ordinary differential equations. The scalar "time" variable t is replaced by several independent variables, say and y, and it is by no means necessary that just one of these should play the role of time in determining the relationships of causality. This opens a great variety of possibilities, as will be shown by the following example.

A GAME WITH THE LAPLACE EQUATION. Let w(x,y) be determined in the rectangle $D:\{(x,y)\,|\,0 \leqq x \leqq 1, 0 \leqq y \leqq 1\}$ by

$$w_{xx}(x,y) + w_{yy}(x,y) = [u(x,y) + v(x,y)]^2,$$

with boundary conditions

$$w(0,y) = w(1,y) = w(x,0) = w(x,1) = 0.$$

Admissible controls are all piecewise continuous functions u(x,y), v(x,y) satisfying $|u(x,y)| \leqq 1$, $|v(x,y)| \leqq 1$ for all (x,y) in D. Let the payoff be

$$J = \iint_D w(x,y)\,dx\,dy,$$

"u" being the minimizing and "v" the maximizing player.

With the aid of the Green's function $G(\xi,\eta,x,y)$ the solution of the equation can be written as

$$w(x,y) = -\iint_D G(\xi,\eta,x,y)\,[u(\xi,\eta) + v(\xi,\eta)]^2\,d\xi\,d\eta.$$

The question which is still crucial for the solution of the game is: what is the sequence in which both players must choose the values of $u(x,y)$ and $v(x,y)$? Or, even more general: what is the information available to each player when he has to choose his control at the point (x,y)?

The game under consideration may, for example, be defined by:

a) When player "u" chooses $u(x,y)$, he knows his opponent's choice of $v(\xi,\eta)$ for all pairs (ξ,η) such that $\xi<y$.

b) When player "v" chooses $v(x,y)$, he knows his opponent's choice of $u(\xi,\eta)$ for all pairs (ξ,η) such that $\eta<x$.

This means that at "time" t ($0 \leqq t \leqq 1$), "u" and "v" have to choose their controls on the lines $y=t$ and $x=t$ respectively, having already chosen the values for $y<t$ and $x<t$ (areas indicated in figure 1).

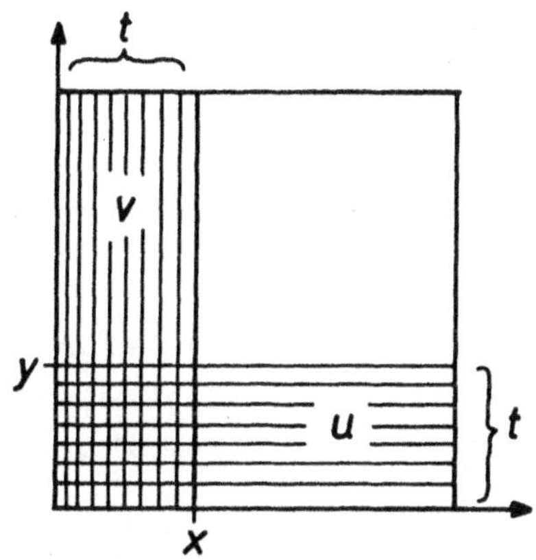

Fig. 1

This game is actually so simple that it can be solved "by inspection". At each point (x,y), player "u" would like to choose $u(x,y) = \operatorname{sgn} v(x,y)$, as $[u+v]^2 = \Delta w = \max!$ leads to $w(x,y)$ minimum. On the other hand, player "v" would like to choose $v(x,y) = -u(x,y)$. This leads to the following optimal strategies:

u-strategy:

$$\text{for } x<y \text{ take } u(x,y) = \begin{cases} \operatorname{sgn} v(x,y) & \text{if } v(x,y) \neq 0 \\ 1 \text{ (or } -1) & \text{if } v(x,y) = 0 \end{cases}$$

for $x>y$ take $u(x,y)$ = any value in [0,1];

v-strategy:

for $x<y$ take $v(x,y) = -\,u(x,y)$

for $x>y$ take $v(x,y) = 0$.

Remark: it is interesting to notice that this game can also be formulated in terms of the variables x,t, where now t denotes "time", as follows:

$$w_{xx}(x,t) + w_{tt}(x,t) = [u(x,t) + v(t,x)]^2.$$

3. EXISTENCE THEORY

For optimal control problems with partial differential equations ("distributed parameter systems") a general theory involving functional analytic methods has been developed (Fattorini [2], A.Friedman [3], Lions [9], Warga [13]).

For differential games of the evolution type

$$w_t = A\,w + B\,u + C\,v,$$

where A is a differential operator considered as an abstract operator in some Banach space, A.Friedman applied his method of defining differential games and showed existence of a solution for some specific type of problems ([4]).

It is not the intention of this paper to go into such technically quite involved considerations. Instead, the naive approach will be followed of trying to find constructive methods for obtaining solutions and, by examples, to find peculiar types of behavior of solutions.

4. ISAACS' EQUATION BY DISCRETIZATION

Isaacs' condition of optimality is the counterpart of what is often referred to as Bellman's "Principle of Optimality" or Pontriagin's "Maximum Principle" in optimal control theory. Applied to systems with partial differential equations, this principle has been established by Butkovskiy [1] and Lions [9] under rather restricted conditions. It will be shown next that Isaacs' equation for partial differential games can be derived from the one for ordinary differential games via discretization of the differential equation. The assumption will be that the physical system in question can be equally well described by a continuous partial differential equation or by a discrete system of sufficiently many ordinary differential equations. By this "elementary" approach, the main and difficult question of giving precise conditions for the validity of Isaacs' equation, is actually avoided, and roughly it is shown that the optimality condition for partial differential games is as good as the optimality condition for ordinary differential games.

A GAME WITH THE HEAT EQUATION. Consider $w(x,t)$ defined in $x \in [0,1]$, $t \in [0,T]$ satisfying

(1) $$w_t = a\, w_{xx} + b\, w_x + c\, w + d + r\, u + s\, v$$

and the boundary conditions

(2) $$w(0,t) = w(1,t) = 0.$$

Here a, b, c, d, r, and s are given functions of (x,t) and $u(x,t)$ and $v(x,t)$ are the controls, for whose determination optimal strategies should be found. Let the payoff be

(3) $$J = \int_0^T \int_0^1 g(x,t,w,u,v)\, dx\, dt,$$

"u" being the minimizer and "v" the maximizer.

In many applications this is the model for a process which can equally well be described by actions on discrete locations. Hence denote by:

x^i, $i=0,1,2,\cdots,N+1$ the points in $[0,1]$ where $w(x,t)$ is observed,

$h = x^{i+1} - x^i$, $x^0 = 0$, $x^{N+1} = 1$,

$w^i = w^i(t) = w(x^i,t)$,

$a^i = a^i(t) = a(x^i,t)$ and similarly for $b^i, c^i, d^i, r^i, s^i, u^i, v^i$;

$g^i(t,w^i,u^i,v^i) = g(x^i,t,w^i(t),u^i(t),v^i(t))$.

Then, with the usual approximations

$$w_x(x^i,t) = \frac{1}{2h}(w^{i+1}(t) - w^{i-1}(t))$$

$$w_{xx}(x^i,t) = \frac{1}{h^2}(w^{i+1}(t) - 2\, w^i(t) + w^{i-1}(t)),$$

the discretized version of (1), (2), (3) becomes

(1d) $$\dot{w}^i = a^i \frac{1}{h^2}(w^{i+1} - 2\, w^i + w^{i-1}) + b^i \frac{1}{2h}(w^{i+1} - w^{i-1}) + + c^i\, w^i + d^i + r^i\, u^i + s^i\, v^i, \qquad (i = 1,\cdots,N)$$

(2d) $$w^0 = w^{N+1} = 0,$$

(3d) $$\bar{J} = h \int_0^T \sum_1^N g^i(t,w^i,u^i,v^i)\, dt.$$

As this is a game defined by a system of ordinary differential equations, Isaacs' conditions of optimality can be applied under the usual assumptions. Hence this condition is

(4d) $$\bar{H} = \sum_1^N p^i\, \dot{w}^i + \sum_1^N g^i = \min_{u^i} \max_{v^i} !$$

Here $\dot{w}^i$ are given by (1d) and p^i $(i = 0, 1,\cdots, N+1)$ are the adjoint state variables, $p^0 = p^{N+1} = 0$. Furthermore,

(5d) $$\dot{p}^i = -\frac{\partial \bar{H}}{\partial w^i} = -\left(\frac{a^{i-1}}{h^2} + \frac{b^{i-1}}{2h}\right) p^{i-1} - \left(-\frac{2a^i}{h^2} + c^i\right) p^i - - \left(\frac{a^{i+1}}{h^2} - \frac{b^{i+1}}{2h}\right) p^{i+1} - \frac{\partial g^i(t,w^i,u^i,v^i)}{\partial w^i}.$$

Going back to the continuous model, equations (4d) and (5d) become

(4) $$H = \int_0^1 [p(x,t)\, w_t(x,t) + g(x,t,u(x,t),v(x,t))]\, dx = \min_{u(x,t)} \max_{v(x,t)} !$$

for each $t \in [0,T]$,

(5) $$p_t(x,t) = -\frac{\partial^2}{\partial x^2}[a(x,t)\, p(x,t)] + \frac{\partial}{\partial x}[b(x,t)\, p(x,t)] - - c(x,t)\, p(x,t) - \frac{\partial g(x,t,w,u,v)}{\partial w}$$

and

(6) $$p(0,t) = p(1,t) = 0.$$

These are the adjoint partial differential equations of the homogeneous part of (1), plus the $\partial g/\partial w$ term.

While this is no proof of the validity of Isaacs' equations as conditions for optimality for games with partial differential equations, it shows that at least for models where discretization is physically meaningful, these equations are equivalent to the corresponding conditions for ordinary differential equations.

5. REDUCTION TO ORDINARY DIFFERENTIAL GAMES BY FOURIER EXPANSION

Consider the game defined by the heat equation

(7) $$w_t(x,t) = w_{xx}(x,t) + y(x,t) + u(x,t) + v(x,t)$$

for $x \in [0,1]$, $t \in [0,T]$, with initial and boundary conditions

(8) $$w(x,0) = 0, \quad w(0,t) = w(1,t) = 0.$$

Here $y(x,t)$ is a given function and $u(x,t)$ and $v(x,t)$ are the controls.

Let the payoff be

(9) $$J = \int_0^1 (w(x,T) - z(x))^2 \, dx$$

for a given end-time $T > 0$. Let "u" be the minimizer and "v" be the maximizer.

Consider the eigenvalues λ_n and eigenfunctions $\phi_n(x)$ of the problem

(10) $$\phi_n'' = \lambda_n \phi_n, \quad \phi_n(0) = \phi_n(1) = 0,$$

which are

$$\lambda_n = -n^2 \pi^2, \quad \phi_n(x) = \sqrt{2} \sin n\pi x.$$

There will be Fourier expansions

$$w(x,t) = \Sigma\, w_n(t)\, \phi_n(x)$$

$$y(x,t) = \Sigma\, y_n(t)\, \phi_n(x)$$

$$z(x) = \Sigma\, z_n\, \phi_n(x)$$

$$u(x,t) = \Sigma\, u_n(t)\, \phi_n(x)$$

$$v(x,t) = \Sigma\, v_n(t)\, \phi_n(x).$$

Assume that the controls are restricted by constraints of the type

$$(11)\qquad \begin{cases} \int_0^1\int_0^1 a(\xi,\eta)\, u(\xi,t)\, u(\eta,t)\, d\xi\, d\eta \leqq 1 \\ \int_0^1\int_0^1 b(\xi,\eta)\, v(\xi,t)\, v(\eta,t)\, d\xi\, d\eta \leqq 1 \end{cases} \quad \text{for each } t.$$

Writing

$$a(\xi,\eta) = \Sigma\, a_n\, \phi_n(\xi)\, \phi_n(\eta)$$

$$b(\xi,\eta) = \Sigma\, b_n\, \phi_n(\xi)\, \phi_n(\eta),$$

this leads to constraints in terms of the Fourier coefficients

$$(12)\qquad \begin{cases} \Sigma\, a_n\, u_n^2(t) \leqq 1 \\ \Sigma\, b_n\, v_n^2(t) \leqq 1 \end{cases} \quad \text{for each } t.$$

It is meaningful to assume that (12) are positive definite and therefore a_n, $b_n > 0$. There may also be a restriction allowing only a certain finite number of coefficients a_n, b_n to be different from zero, condition which would delete all problems concerning convergence of these series.

In terms of the Fourier coefficients, equation (7) becomes

$$(13)\qquad \dot{w}_n(t) = -\, n^2\, \pi^2\, w_n(t) + y_n(t) + u_n(t) + v_n(t),$$

with the initial condition

$$(14)\qquad w_n(0) = 0$$

and payoff

$$(15)\qquad J = \Sigma\, (w_n(T) - z_n)^2.$$

Applying the Isaacs' equation to the game (13), (14), (15), the following conditions are obtained for optimality:

$$(16)\qquad H = \Sigma\, p_n\, \dot{w}_n = \Sigma\, p_n\, [\, -n^2\pi^2 w_n + y_n + u_n + v_n] = \min_{u_n} \max_{v_n} !,$$

$$(17)\qquad \dot{p}_n(t) = n^2\, \pi^2\, p_n(t),$$

from which

$$p_n(t) = p_n(T)\, e^{-n^2\pi^2(T-t)}$$

follows. Hence, according to condition (16), $u_n(t)$ and $v_n(t)$ should be optimally determined for each t such that

$$(18)\qquad \begin{cases} \Sigma\, p_n(t)\, u_n(t) = \Sigma\, p_n(T)\, e^{-n^2\pi^2(T-t)}\, u_n(t) = \min! \\ \Sigma\, p_n(t)\, v_n(t) = \Sigma\, p_n(T)\, e^{-n^2\pi^2(T-t)}\, v_n(t) = \max! \end{cases}$$

subject to the constraints (12).

Using Lagrange multipliers λ, μ, this can be written as

$$\Sigma\, p_n(T)\, e^{-n^2\pi^2(T-t)}\, u_n(t) - \lambda(t)\, \Sigma\, a_n\, u_n^{\,2}(t) = \text{min!}$$

$$\Sigma\, p_n(T)\, e^{-n^2\pi^2(T-t)}\, v_n(t) - \mu(t)\, \Sigma\, b_n\, v_n^{\,2}(t) = \text{max!}$$

Hence, the optimal controls should be determined from

$$(19) \qquad \begin{cases} u_n(t) = \dfrac{p_n(T)\, e^{-n^2\pi^2(T-t)}}{2\,\lambda(t)\, a_n} \\[2ex] v_n(t) = \dfrac{p_n(T)\, e^{-n^2\pi^2(T-t)}}{2\,\mu(t)\, b_n} \end{cases}$$

subject to (12).

Discussion. The values of $p_n(T)$ are in general unknown. They depend on the end-state $w(x,T)$ (or the $w_n(T)$'s) reached after optimal play. Their values represent the components of the (infinite-dimensional) gradient of the "Value" of the game, which according to (9) is

$$V = J_{opt} = V(w(T)) = \Sigma\, (w_n(T) - z_n)^2.$$

Therefore

$$(20) \qquad p_n(T) = \frac{\partial V}{\partial w_n} = 2\,(w_n(T) - z_n).$$

From the end-states, the optimal paths can be constructed backwards as usual ("Retrogression Principle" of Isaacs' [7]).

If, at the end, $w_n(T) = z_n$ for all n, i.e. if the desired goal-function $z(x)$ is actually attained, then all $p_n(T) = 0$ according to (20) and the solution is not determined by these equations. This means that such a state can be optimally reached from several directions in the Banach space.

Another remark is that

$$\frac{u_m(t)}{u_n(t)} = k_{m,n}\, e^{-(m^2-n^2)\pi^2(T-t)} \to 0$$

as $t \to -\infty$ for $m > n$, and similarly for $v_m(t)/v_n(t)$. Hence the higher harmonics in the controls become less and less important as the time $(T-t)$ to finish the game gets larger.

A final remark is that this development is rather formal, in the sense that no precise conditions for convergence nor optimality were formulated. For example, in establishing (16) and (17), the well known Isaacs' theory was used, but this theory is usually given only for finite-dimensional systems, while the present one is infinite-dimensional. The results given here can therefore only be expected to give the true solution if convergence of the series is strong enough.

6. BOUNDARY CONTROLLED HEAT EQUATION

Consider the game defined by the heat equation

$$w_t(x,t) = w_{xx}(x,t), \tag{21}$$

where $x \in [0,1]$, $t \in [0,T]$, with initial condition

$$w(x,0) = w_0(x) \tag{22}$$

and boundary controls

$$\begin{aligned} w(0,t) &= u(t), \\ w(1,t) &= v(t), \end{aligned} \tag{23}$$

subject to the constraints

$$\begin{aligned} |u(t)| &\leqq M_1 \\ |v(t)| &\leqq M_2. \end{aligned} \tag{24}$$

Let the payoff be

$$J = \int_0^1 |w(x,T)|\ dx \tag{25}$$

for some prescribed value of $T > 0$. Player "u" should minimize and player "v" should maximize this payoff.

This game was described by Roxin in [11], and gives the opportunity to discuss interesting problems related to the solutions. Boundary controls in higher dimensional heat equation were considered recently by Glashoff and Weck ([5]).

In this game, the optimal controls may present a denumerable infinity of jumps (see Butkovskiy [1], where this arises in optimal control problems), all in the neighborhood of the end-time T. In order to avoid this, the condition

$$u(t) = v(t) = 0 \quad \text{for} \quad T-\varepsilon < t \leqq T \tag{26}$$

will be imposed, for a prescribed value of $\varepsilon > 0$.

The solution of equations (21), (22), (23) is given by

$$w(x,t) = w_0(x,t) + w_1(x,t) + w_2(x,t), \tag{27}$$

where w_0 arises from the initial condition and w_1 and w_2 from the controls. Their Fourier expansions in terms of the eigenfunctions (10) is

$$w_0(x,t) = \Sigma\ e^{-n^2\pi^2 t}\ \phi_n(x) \int_0^1 w_0(\xi)\ \phi_n(\xi)\ d\xi, \tag{28}$$

$$w_1(x,t) = \sqrt{2}\,\pi\,\Sigma\, n\,\phi_n(x) \int_0^t e^{-n^2\pi^2(t-\tau)}\, u(\tau)\, d\tau,$$

(29)

$$w_2(x,t) = \sqrt{2}\,\pi\,\Sigma\,(-1)^{n+1}\, n\,\phi_n(x) \int_0^t e^{-n^2\pi^2(t-\tau)}\, v(\tau)\, d\tau.$$

These can be written also as

(30) $$w_0(x,t) = \int_0^1 k(x,t,\xi)\, w_0(\xi)\, d\xi$$

and

$$w_1(x,t) = \int_0^t K_1(x,t-\tau)\, u(\tau)\, d\tau$$

(31)

$$w_2(x,t) = \int_0^t K_2(x,t-\tau)\, v(\tau)\, d\tau,$$

where

(a) $$k(x,t,\xi) = \Sigma\,\phi_n(x)\,\phi_n(\xi)\, e^{-n^2\pi^2 t}$$

(32) (b) $$K_1(x,t-\tau) = \sqrt{2}\,\pi\,\Sigma\, n\,\phi_n(x)\, e^{-n^2\pi^2(t-\tau)}$$

(c) $$K_2(x,t-\tau) = \sqrt{2}\,\pi\,\Sigma\,(-1)^{n+1}\, n\,\phi_n(x)\, e^{-n^2\pi^2(t-\tau)}.$$

It should be remarked that (b) and (c) have singularities at $\tau = t$, hence the convenience of having the restriction (26).

From (27), (30) and (31) it follows that the payoff, expressed in terms of the controls, is

(33) $$J = \int_0^1 \Big| w_0(x,T) + \int_0^T K_1(x,T-\tau)\, u(\tau)\, d\tau + \int_0^T K_2(x,T-\tau)\, v(\tau)\, d\tau \Big|\, dx.$$

Let $w^*(x,t)$ be the optimal solution, $u^*(t)$ and $v^*(t)$ the optimal controls and J^* the optimal payoff. Define the sets

(34)

$$P = \{ x \in [0,1] \mid w^*(x,T) > 0 \}$$

$$N = \{ x \in [0,1] \mid w^*(x,T) < 0 \}.$$

Then (33) can be written as

(35) $$J^* = \Big\{ \int_P - \int_N \Big\}\Big[w_0(x,T) + \int_0^T K_1(x,T-\tau)\, u^*(\tau)\, d\tau + \int_0^T K_2(x,T-\tau)\, v^*(\tau)\, d\tau \Big]\, dx,$$

where the shorthand notation

$$\Big\{ \int_P - \int_N \Big\} F(x)\, dx = \int_P F(x)\, dx - \int_N F(x)\, dx$$

has been used.

By interchanging the order of integration, (34) becomes

$$(36)\quad \begin{aligned} J^* &= \{\int_P - \int_N\} w_0(x,T)\, dx + \\ &+ \int_0^T u^*(\tau)\, d\tau \{\int_P - \int_N\} K_1(x,T-\tau)\, dx + \\ &+ \int_0^T v^*(\tau)\, d\tau \{\int_P - \int_N\} K_2(x,T-\tau)\, dx. \end{aligned}$$

For optimality of the controls, this expression of J* should be compared with other values of J, for which the sets P, N would be different. Nevertheless, it is not difficult to see that in "linear approximation", P and N may be taken as constant, obtaining

$$\begin{aligned} \delta J &= \int_0^T \delta u(\tau) \{\int_P - \int_N\} K_1(x,T-\tau)\, dx + \\ &+ \int_0^T \delta v(\tau) \{\int_P - \int_N\} K_2(x,T-\tau)\, dx . \end{aligned}$$

It follows immediately that

$$(37)\quad \begin{aligned} u^*(t) &= - M_1 \operatorname{sgn} \{\int_P - \int_N\} K_1(x,T-\tau)\, dx \\ v^*(t) &= M_2 \operatorname{sgn} \{\int_P - \int_N\} K_2(x,T-\tau)\, dx \end{aligned}$$

form an open-loop Nash equilibrium point in the class of neighborhood controls of (u*,v*).

Equations (37) work according to the "Principle of Retrogression" of Isaacs ([7]). Indeed, one should know the optimal end-state w*(x,T) in (34), in order to be able to construct the optimal solution backwards. But there are more problems concerning this backwards construction of the solution.

Starting from an arbitrary end-state w*(x,T), two main questions arise:

a) How far backwards in time does a solution satisfying the optimality conditions (37) exist?

b) How far backwards is this solution really optimal?

The second question also appears naturally in many ordinary differential games, where optimal solutions may converge (or diverge) on "singular manifolds" (particularly "dispersal manifolds").

The first question is very important for the heat equation (21), as it is well known that the initial-value problem for this equation becomes ill-posed if the time is reversed.

A related interesting question, not necessarily concerned with optimality, is the following. Given an end-state $w(x,T)$, equation (21) can usually be integrated backwards in time with boundary values $w(0,t) = w(1,t) = 0$ only for a finite time interval $\overline{T} < t \leqq T$. At $t = \overline{T}$, indeed, the Fourier-series solution

$$w(x,t) = \Sigma\, e^{n^2\pi^2(T-t)}\, \phi(x) \int_0^1 w(\xi,T)\, \phi_n(\xi)\, d\xi$$

will become divergent. Question: under what conditions is it possible to find boundary controls $w(0,t)$ and $w(1,t)$ such that the interval of existence of the solution will be extended backwards beyond $\overline{T}$? How far backwards can the solution be extended in this way?

7. BOUNDARY CONTROLLED WAVE EQUATION

Consider the wave equation

(38) $$w_{tt}(x,t) = w_{xx}(x,t)$$

in $x \in [0,1]$, $t \leqq 0$. The concept of solution adopted here will be that of a piecewise smooth function; edges and even discontinuities will be admitted if they arise from similar properties of the boundary values.

Let the initial conditions be

(39) $$w(x,0) = \phi(x), \quad w_t(x,0) = \psi(x).$$

The boundary controls

(40) $$w(0,t) = u(t), \qquad w(1,t) = v(t)$$

will be restricted by

(41) $$|u(t)| \leqq M_1 \ , \ |v(t)| \leqq M_2.$$

Differential games of this type were considered by Roxin in [10], [12].

The well known form of the solution is

(42) $$w(x,t) = w_0(x,t) + w_1(x,t) + w_2(x,t),$$

where

(43) $$w_0(x,t) = \tfrac{1}{2}[\phi(x-t) + \phi(x+t) + \int_{x-t}^{x+t} \psi(s)\, ds]$$

is the part of the solution generated by the initial conditions (39). In this formula the functions ϕ and ψ should be extended as odd functions, periodic of period 2. The boundary conditions (40) generate

(44) $$\begin{aligned} w_1(x,t) &= u(t-x) - u(t+x-2) + u(t-x-2) - u(t+x-4) + \cdots \\ w_2(x,t) &= v(t-1+x) - v(t-1-x) + v(t-3+x) - v(t-3-x) + \cdots . \end{aligned}$$

Let the terminal time be the first time $T > 0$ such that a certain condition is satisfied, and let this "terminal condition" be

$$\int_0^1 w(x,T)\,dx = K, \tag{45}$$

where K is a given real number. Let the payoff be

$$J = T, \tag{46}$$

"u" being the minimizer and "v" the maximizer. This is terefore a "time-optimal" (or "survival") game.

In spite of the fact that the governing equations (38) and (45) are linear, the payoff (46) is nonlinear in the controls and the solution of this game is not trivial.

Let

$$M_1 > M_2 > 0, \tag{47}$$

let $u^*(t)$ and $v^*(t)$ be the controls arising from optimal play, and let $w^*(x,t)$ and T^* be the corresponding state-function and termination time.

From formula (44) it follows that the boundary values $u(t)$, $v(t)$ propagate in the interior of the unit x-interval along the characteristic lines, being reflected at the boundaries $x = 0$ and $x = 1$ with change of sign. Once T^* is known, the optimal boundary values are easily constructed in reversed time-order. Indeed, assuming that the terminal condition $\int w\,dx = K$ is reached from the side of smaller values of this integral, "u" will be interested in making this integral as big as possible, and hence will plan his boundary values in such a way that after suitable reflections they arrive at the end-line $t = T$ as positive M_1; "v" will face a similar situation but with reverse sign. In figure 2 such optimal boundary values are indicated for $\frac{5}{2} < T^* < 3$.

At this point there still remains the question of how to determine the value T^*.

In case of zero initial conditions $w(x,0) = w_t(x,0) = 0$, it is easy to find with the aid of the figure 2, that

$$\int_0^1 w(x,T^*)\,dx = (M_1 - M_2)\,T^*.$$

Hence, in this case $T^* = K/(M_1 - M_2)$.

For non-zero initial conditions (39), superposition gives

$$\int_0^1 w(x,T^*)\,dx = (M_1 - M_2)\,T^* + I_0(T^*), \tag{48}$$

where $I_0(t)$ is the value of the integral for zero boundary conditions, which can be calculated in advance as it does not depend on the controls.

In order to calculate $I_0(t)$, the solution $w_0(x,t)$ can be written as a superposition of two waves

$$w_0(x,t) = F(x-t) + G(x+t),$$

traveling in the positive respectively negative x-direction, and these waves get reflected at the boundaries with change of sign. Hence

$$I_0(t) = \int_0^1 w_0(x,t)\, dx$$

is a periodic function satisfying $I_0(t+2) = - I_0(t+1) = I_0(t)$.

The terminal time T* will therefore be determined in this case by the condition

$$(M_1 - M_2)\, T^* + I_0(T^*) = K \tag{49}$$

and will be the smallest value of T* satisfying this condition. The solution of this equation leads to a graph as represented in figure 3.

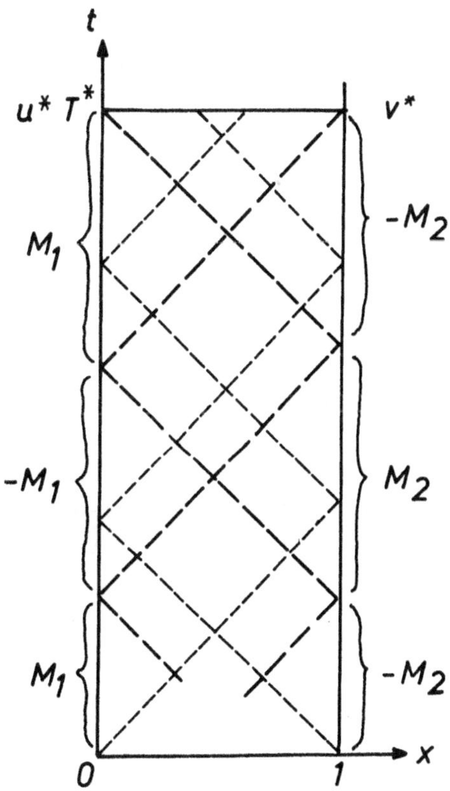

Fig. 2

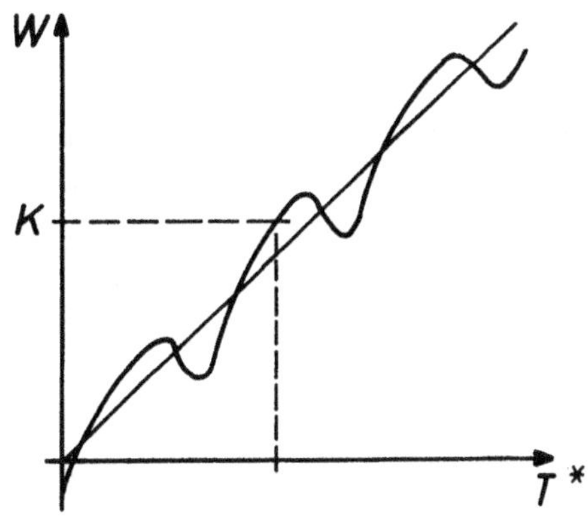

Fig. 3

USEABLE PART OF THE TERMINAL MANIFOLD. The "terminal manifold" is the set of pairs (w,t) satisfying the terminal condition (45) (here w should be considered as a function w(x), element of some Banach space, depending on a parameter t). It is important to determine the "useable" part of this manifold, i.e. the set of pairs (w,t) which are really terminal for a possible optimal play (Isaacs, [7]).

Using the notation

$$W(t) = \int_0^1 w(x,t)\,dx \tag{50}$$

and assuming that at the initial time, $W(t_0) < K$, the terminal manifold will be reachedby increasing values of W(t). Hence the useable part is characterized by the fact that, under optimal play, $\frac{d}{dt}W(t) > 0$.

A pair (w,t) on the terminal manifold will be useable if considered as initial condition (hence given by ϕ, ψ according to (39)),

$$\max_u \min_v [dW(t)/dt] > 0. \tag{51}$$

In order to calculate this derivative, consider separately, as in (42),

$$\begin{aligned} W(t) &= W_0(t) + W_1(t) + W_2(t) = \\ &= \int w_0\,dx + \int w_1\,dx + \int w_2\,dx. \end{aligned} \tag{52}$$

Here

$$\begin{aligned} W_0(t) &= \frac{1}{2}\int_0^1 [\phi(x+t) + \phi(x-t)]\,dx + \\ &+ \frac{1}{2}\int_0^1 dx \int_{x-t}^{x+t} \psi(s)\,ds. \end{aligned} \tag{53}$$

If it is assumed that $\psi(x)$ is continuous in (0,1), having possibly discontinuities only at x=0 and x=1 (due to the odd-periodic continuation; this may not be the case if prior boundary values were discontinuous), then one finds that the derivative $d/dt|_{t=0}$ of the first term in (53) is $-[\phi(0^+) + \phi(1^-)]$, while the corresponding derivative of the second term is $\int_0^1 \psi(x)\,dx$. Hence

$$\left.\frac{dW_0(t)}{dt}\right|_{t=0} = -[\phi(0^+) + \phi(1^-)] + \int_0^1 \psi(x)\,dx.$$

It is also easy to find

$$\left.\frac{dW_1(t)}{dt}\right|_{t=0} = u(0), \qquad \left.\frac{dW_2(t)}{dt}\right|_{t=0} = v(0).$$

Hence, according to (51), the condition

$$M_1 - M_2 - [\phi(0^+) + \phi(1^-)] + \int_0^1 \psi(x)\,dx > 0 \tag{54}$$

characterizes the useable part of the terminal manifold. Of course,

$$\int_0^1 \phi(x)\, dx = K$$

should also be satisfied.

REMARK ABOUT THE OPTIMAL STRATEGIES. The optimal strategies for this game, leading to the optimal controls, can be stated as follows:

"u" strategy: first determine T^* by condition (49), as explained above; then play the corresponding $u^*(t)$ open loop.

"v" strategy: play constantly $v(t) = - u(t)$.

It is indeed easy to check that these strategies constitute a Nash equilibrium point.

Strategies like the "v" strategy given above, which depend on the instantaneous choice of the opponent's control, are called "stroboscopic" according to Hajek ([6]). While its practical implementation is not possible in an absolutely precise way, in many cases a good approximation with a small delay may be almost as good as the theoretically optimal control.

REFERENCES

[1] A. G. Butkovskiy, Distributed Control Systems, Mod.Anal.Comput. Methods in Sci. and Math., Elsevier, New York, 1969.

[2] H. O. Fattorini, Control in Finite Time of Differential Equations in Banach Space, Commun.Pure Appl.Math.19 (1965), 17-34.

[3] A. Friedman, Optimal Control in Banach Space with fixed End Points, J.Math.Anal.Appl.24 (1968), 161-181.

[4] A. Friedman, Differential Games, Wiley-Interscience, New York 1971.

[5] K. Glashoff and N. Weck, Boundary Control of Parabolic Differential Equations; Sup.Norm Problems, SIAM J.Control Optim. 14 (1976), 662-681.

[6] O. Hajek, Pursuit Games, Math.Sci.Engr., Academic Press, New York 1975.

[7] R. Isaacs, Differential Games, J.Wiley & Sons, New York 1965.

[8] V. Komkov, Optimal Control Theory for the Damping of Vibrations of Simple Elastic Systems, Lect.Notes in Math. 253, Springer Berlin, Heidelberg, New York 1972.

[9] J. L. Lions, Contrôle Optimal de Systèmes Gouvernés par des Équations aux Dérivées Partielles, Dunod-Gauthier Villars, Paris 1968.

[10] E.Roxin, Differential Games with Partial Differential Equations, in Theory and Appl. of Differential Games, Proc. NATO Adv. Study Inst. Warwick, England, edit. J.D.Grote, D.Reidel Publ. Co., Dordrecht, Boston 1975.

[11] E.Roxin, Dispersal Manifolds in Partial Differential Games, to be publ. Proc. Intern. Symp. Dynam.Systems, Gainesville,Fla.,USA, 1975.

[12] E.Roxin, Singular Manifolds in Partial Differential Games, to be publ. Proc.Second Kingston Conf.Differential Games and Opt. Control, Kingston,RI,USA, Marcel Dekker, 1976.

[13] J. Warga, Optimal Control of Differential and Functional Equations, Academic Press, New York, 1972.

COLLISION AVOIDANCE AT SEA

Thomas L. Vincent

Aerospace and Mechanical Engineering
University of Arizona, Tucson, 85721/USA

ABSTRACT: The problem of collision avoidance between surface ships may be examined from the perspective of either macro or micro collision avoidance. In macro collision avoidance the pilot of a ship would try for maintaining a given mileage separation between his own ship and others. Micro collision avoidance would be concerned with manoeuvres between ships of less than a given separation distance. Analysis of the two situations is different. In macro collision avoidance the other ships would often be maintaining constant speed and heading. Attention would be focused on keeping other ships out of an appropriate circle about the center of ones own ship. In micro collision avoidance other ships would in general be manoeuvering. The actual shape of the ships may be important in determining appropriate manoeuvers for own ship. In both macro and micro collision avoidance, qualitative game/control theory is useful for the determination of critical barriers and manoeuvres. Results obtained this way for two ships are compared with some collision avoidance guidelines in actual use.

INTRODUCTION

Today, the threat of collision at sea is of direct concern not only to the mariner and his passengers, but to any one of environmental inclination as well. According to Webster (1974) "In 1970 one out of every fourteen ships of 500 gross tons or more was involved in a collision ... one out of twenty-five resulted in total loss ... that is, in 1970, one out of 350 large ships sailing the ocean was completely lost." Evidently 1970 was not a particularly bad year, Miloh (1974) states, "... in 1972 approximately 3000 ships over 500 gross registered tons were involved in collisions ... one out of forty resulted in a total loss."

On the surface these facts are difficult to reconcile. After all, since it is difficult to rapidly change the speed of a large ship, most manoeuvres are seemingly simple and secondly, manoeuvres must comply with official rules of the nautical road with a history dating from 1895. Obviously numerous collisions do occur in spite of (or perhaps because of) these factors. Indeed, observed manoeuvres are not simple to understand when viewed from the relative reference of a ship in motion and because of legalese, the rules of the road are often obscure [Webster (1974)].

According to the rules, in a risk of collision situation between two ships, either both ships are *burdened* to manoeuvre, or one is burdened while the other is *privileged* to maintain course and speed. Should the situation get more dangerous, such that the ships are in *extremis*, then both ships are always burdened to evade each other.

This suggests that collision avoidance be implemented at two levels. One might call them macro and micro collision avoidance. In macro collision avoidance the pilot of a burdened ship should try for maintaining a separation distance between his *own* ship and any other *target* ship no less than that considered proper for safe passage. Micro collision avoidance would be associated with manoeuvres between ships of less than this separation distance.

In macro collision avoidance, an appropriate *safe passage circle* (SPC) is envisioned about the center of own ship (considered burdened) and attention is directed toward manoeuvring to keep one or more potential targets out of the SPC. With only two ships under constant speed, macro collision avoidance is related to Isaacs (1965) game of two cars. A qualitative game/control theoretic approach to this problem has been examined previously by Vincent *et al.* (1972) in the context of aircraft collision avoidance and by Miloh (1974), and Miloh and Sharma (1975) in the context of ship collision avoidance.

Should a target enter the SPC then the ships would be in extremis and the danger of collision is deemed sufficiently high as to require a fine tuning of manoeuvres. Thus with micro collision avoidance, the actual shape of the ships may become important in determining appropriate manoeuvres. In this case manoeuvering is for the avoidance of actual physical contact. A qualitative game/control theoretic approach to micro collision avoidance has been previously examined by Vincent and Peng (1973) and Vincent (1975). A geometric approach to this same problem is given by Webster (1974).

THE PROBLEM

The visual or radar motion of a target ship can be viewed either with respect to a fixed direction in space or with respect to a coordinate system relative to own ship. If the coordinate system is fixed to own ship, then any drift by own ship will appear as a drift in the target.

This complicates the visual perception of target motion and hence the risk of collision. For this reason the experienced mariner fixes visual targets against a *stabilized* reference such as stars or compass direction when making an assessment of target motion. In order for this same mariner to use his visual experiences with a radar display as a source of information, a stabilized radar should be used.

The equations governing the motion of a target *blip* on a stabilized radar screen are given by

$$\dot{x}_1 = s_t \sin \gamma_t - s_0 \sin \gamma_0 \quad (1)$$

$$\dot{x}_2 = s_t \cos \gamma_t - s_0 \cos \gamma_0 \quad (2)$$

where the dot denotes a time derivative and x_1 and x_2 are the coordinates of the target with respect to a non-rotating Cartesian coordinate system centered at own ship, γ_t and γ_0 are the heading angles of the target and own ship respectively as measured from the positive x_2 axis (due North), s_t and s_0 are the speeds of the target and own ship respectively.

Since the rate of change of speed for a large ship is small, it will be assumed in all further discussion that s_0 and s_t are constant. Changes in the heading angles are determined by turning right or left at some rate within bounds according to

$$\dot{\gamma}_t = u \quad (3)$$

$$\dot{\gamma}_0 = v \quad (4)$$

where $|u_{max}| \geq u$ and $|v_{max}| \geq v$. It is assumed that u_{max} and v_{max} are functions of ship design and independent of speed. Thus if two ships are identical, the faster ship will have a larger turning radius.

If a ship is equipped with a stabilized radar which has a high persistence phosphar, then the trajectory of a target is automatically plotted by the blip in accordance with equations (1) and (2). For example, figure 1 illustrates the use of this equipment. Own ship heading is indicated by a *heading flasher* (due North in figure 1). The circles indicate mileage from own ship at center. Three targets are shown. Assume that any passage greater than one mile is considered safe. If all ships (including own) maintain current speed and heading then both target 1 and 2 will penetrate the one mile separation distance and target 3 will pass within 2½ miles of own ship. The display yields the relative headings of the targets, but not the time of penetration or passage. If own ship is burdened, the pilot must not only decide what

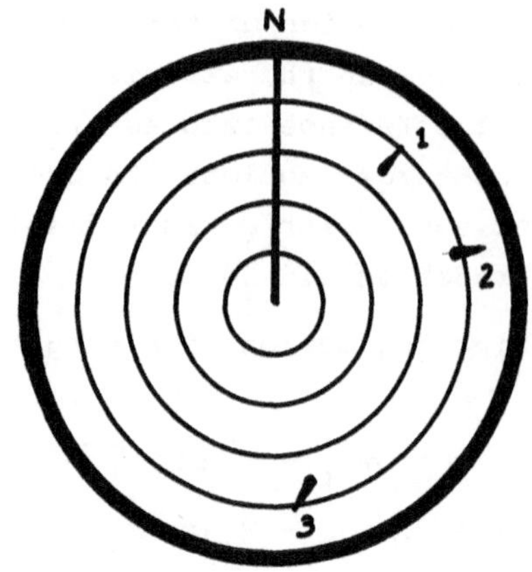

FIGURE 1. TARGET MOTION ON A STABILIZED RADAR

manoeuvres to make, but when to initiate them. An immediate change of heading by own ship, upon discovering that a target is on a collision course, is not necessarily wise. The distance of target from own ship, closing speed, and location of other targets are all factors to be accounted for in making a prudent decision.

Carpenter and Waldo (1974) have obtained a set of rules to aid the navigator in deciding what action to take. These rules were obtained experimentally by using a radar simulator to examine known techniques of collision avoidance. The rules refer to the motion of a target blip under manoeuvres by *own ship only*. Three of these rules have been paraphrased from this reference.

Rule 1. If the angle between the heading flasher and the heading of a target is less than 90° (relative motion is "up-the-scope") then when own ship turns right, the target will appear to turn left.

Rule 2. If the angle between the heading flasher and the heading of a target is greater than 90° (relative motion is "down-the-scope") then when own ship turns right the target will appear to turn right.

Rule 3. If the angle between the heading flasher and the heading of a target is approximately 90° ("being in limbo") then a turn by own ship will initially have little effect on the direction of the target.

Actually the rules follow directly from equations (1) and (2). The dot product between the heading flasher vector $\vec{V} = s_0 \sin\gamma_0 \hat{i} + s_0 \cos\gamma_0 \hat{j}$ and the relative velocity vector $\dot{\vec{V}}_r = \dot{x}_1 \hat{i} + \dot{x}_2 \hat{j}$ ($\hat{i}$ and $\hat{j}$ unit vectors) yields

$$\vec{V}_r \cdot \dot{\vec{V}}_r / s_0 = s_t \cos(\gamma_t - \gamma_0) - s_0 . \tag{5}$$

The cross product between $\vec{V}_r$ and $\dot{\vec{V}}_r = - s_0 \dot{\gamma}_0 \cos\gamma_0 \hat{i} + s_0 \dot{\gamma}_0 \sin\gamma_0 \hat{j}$ (assuming $\dot{\gamma}_t = 0$) yields

$$\vec{V}_r \times \dot{\vec{V}}_r / s_0 = \dot{\gamma}_0 [s_t \cos(\gamma_t - \gamma_0) - s_0]\hat{k} \tag{6}$$

where $\hat{k}$ is a unit vector in the direction of $\hat{i} \times \hat{j}$. From (5) it follows that the angle between the heading flasher and the heading of the target will be less than 90° provided that

$$s_t \cos(\gamma_t - \gamma_0) > s_0 \tag{7}$$

a situation which will never be satisfied unless the target is faster than own ship. The angle between the heading flasher and the heading of the target will be greater than 90° provided that

$$s_t \cos(\gamma_t - \gamma_0) < s_0 \tag{8}$$

a situation which can be satisfied with the target faster or slower than own ship.

Rule 1 follows from (6) when (7) is satisfied. (A positive cross product indicates a left turn by the blip. Note that $\dot{\gamma}_0$ is positive when own ship turns right.) Rule 2 follows from (6) when (8) is satisfied. Thus the rules are sound, and according to Carpenter and Waldo (1974) they are effective and are currently being used at sea by many graduates of the Maryland Maritime Institute of Technology and Graduate Studies.

The rules definitely help in the understanding of relative motion. However there remains the fundamental difficulty associated with rules; rules 1 and 2 can appear to be violated if during a turn $s_t \cos(\gamma_t - \gamma_0) = s_0$. Perhaps the most serious aspect of depending on these rules for any close quarter situation is that they are unreliable if the target ships are also manoeuvring.

It would appear then that the above rules are useful for a macro collision avoidance situation where it is reasonable to assume that all targets are *standing on*, that is maintaining current speeds and headings. However even under this assumption more definitive information on when to make a turn, how much of a heading change to make, and even whether to turn right or left is in order.

Consider now a problem in macro collision avoidance involving only one target. Assume that own ship is burdened and that the target is standing on. If the target is so situated that control action by own ship can keep the target from entering the SPC then there is no real risk of collision. However if the target is situated so that it will ultimately penetrate the SPC regardless of control action by own ship to the contrary, then there is a clear risk of collision. The dividing line between these two cases is of interest as it can be used to identify when avoidance manoeuvres should be initiated. A qualitative control analysis may be used to find such a dividing line.

RISK OF COLLISION

The barrier concepts used in differential games [Isaacs (1965)] have analogs in control systems. Some geometric properties of a particular "control barrier" will now be examined.

Consider a control system of the form

$$dx/dt = f(x,u) \tag{9}$$

where $f(\cdot) : R^n \times R^r \to R^n$ is some specified C^1 function of the state vector x and control vector u. An $n-1$ dimensional set of points M is said to be a *Membrane* if for every $x \in M$ there exists

(i) a ball B centered at x such that the membrane partitions the ball into two non-empty connected n dimensional subsets B^1 and B^2.

(ii) a number $\delta > 0$ such that for any piecewise continuous control function $u(t) \in U \subset R^r$, $t \in [0,\gamma]$, the trajectory obtained by integrating (9) starting from $x(0) \in M$ is an element of B^1 for all $t \in (0,\delta)$.

(iii) a piecewise continuous control function $u^*(t) \in U$, $t \in [0,\delta]$ such that the trajectory $x(t)$ obtained by integrating (9) starting from $x(0) \in M$ is an element of M for all $t \in [0,\delta]$.

Consider now a set defined by

$$\mathbf{I} = \{x \in R^n | \theta_\sigma(x) \le 0, \quad \sigma \in S\} \tag{10}$$

where $\theta_\sigma(\cdot) : R^n \to R^1$ is C^1 on R^n, $S = \{1 \ldots s\}$ and s is a positive integer. Let $S_p = \{\sigma | \theta_\sigma(x^p) = 0\}$, $x^p \in \mathbf{I}$. From tangent vector considerations and the first order approximation theorem, it follows that a sufficient condition for $x(t)$ to be an element of comp $\mathbf{I}$ for all $t \in (0,\delta)$, $\delta > 0$, starting from $x(0) \in \partial\mathbf{I}$ is that there exists a $u \in U$ and a $\sigma \in S_p$ such that

$$\partial\theta_\sigma/\partial x|_{x(0)} \cdot f(x(0),u) > 0. \tag{11}$$

Theorem: If there exists a function $\theta_1(\cdot) : E^n \to R^1$ which is C^1 on R^n such that

$$M = \{x \in E^n | \theta_1(x) = 0\} \tag{12}$$

and either

$$\bar{B}^1 = \{x \in E^n | \theta_1(x) \leq 0, \; |x - x(0)| < r\} \tag{13}$$

or

$$\bar{B}^1 = \{x \in E^n | \theta_1(x) \geq 0, \; |x - x(0)| < r\} \tag{14}$$

where r is the radius of a ball B about $x(0) \in M$. Then there exists a control vector $u^* \in U$ such that either

$$\partial\theta_1/\partial x|_{x(0)} \cdot f(x(0),u) \leq \partial\theta_1/\partial x|_{x(0)} \cdot f(x(0),u^*) = 0 \tag{15}$$

or

$$\partial\theta_1/\partial x|_{x(0)} \cdot f(x(0),u) \geq \partial\theta_1/\partial x|_{x(0)} \cdot f(x(0),u^*) = 0 \tag{16}$$

for all $u \in U$.

Proof: Let $x(t)$ be the trajectory obtained by integrating (9) with the control $u \in U$ starting from a point $x(0) \in M$. By definition of M, $x(t) \in B$, $t \in [0,\delta]$, thus from (11) it follows that for all $u \in U$

$$\partial\theta_1/\partial x|_{x(0)} \cdot f(x(0) \leq 0 \; . \tag{17}$$

By (iii) of the definition the equality follows.

Let $H = \partial\theta_1/\partial x|_{x(0)} \cdot f[x(0),u]$. Then the theorem implies that at every $x \in M$ the function H is either maximized or minimized with respect to u. A given point $x \in R^n$ may be an element of one or more membranes. In general if $x \in M$ such that H is maximized, there is another membrane passing through x such that H is minimized.

Let $\mathbf{I}$ represent the set of points enclosed by an SPC as illustrated in figure 2. Suppose that M_1, M_2, M_3, and M_4 are all the membranes which are support hyperplanes to $\mathbf{I}$ and that the direction of the gradient to each membrane is as indicated. Suppose that H is maximized with respect to u on M_1 and M_2, and that H is minimized with respect to u on M_3 and M_4. A number of regions with distinctly different properties have been delineated by these membranes.

Let $u(\cdot)$ designate a piecewise continuous control program $u(t) \in U$ for all $t \in [0,t_f)$ and let $x(t)$ be the trajectory obtained by integrating (9) with $u(\cdot)$ starting from $x(0)$. Then:

(i) the point x^p is *reachable* from $\mathbf{I}$ if for some $x(0) \in \mathbf{I}$ there exists a $u(\cdot)$ such that $x(t_f) = x^p$ for some $t_f \geq 0$.

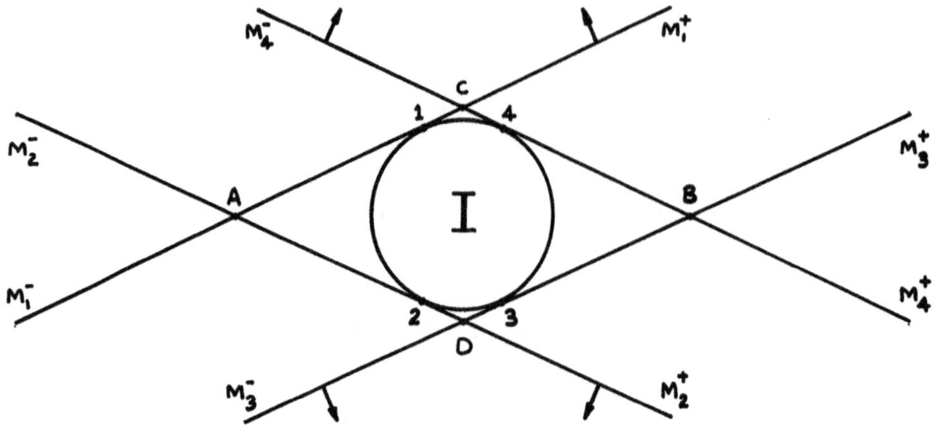

FIGURE 2. MEMBRANES AS SUPPORT HYPERPLANES

(ii) the point x^p is *controllable* to **I** if given $x(0) = x^p$, there exists a $u(\cdot)$ such that $x(t) \in$ **I** for some $t_f \geq 0$.

(iii) the point $x^p \notin$ **I** *collides* with **I** if given $x(0) = x^p$, $t_f = \infty$, and any $u(\cdot)$, $x(t) \in$ **I** for some $t \in (0,t_f)$.

(iv) the point $x^p \notin$ **I** is *shadowed* by **I** if given $x(0) = x^p$, $t_f = \infty$, and any $u(\cdot)$, $x^r(t) \in$ **I** for some $t \in (0,t_f)$ where $x^r(t)$ is the solution to the *retro* equation $\dot{x} = -f(x,u)$.

(v) the point $x^p \notin$ **I** is *ambivalent* with respect to **I**, if for some $x(0) \in$ **I**, there exists a $u(\cdot)$ such that $x(t_1) = x^p$ and $x(t_f) \in$ **I** for some $0 \leq t_1 \leq t_f$.

(vi) etc.

In reference to figure 2 it follows that:

(i) Region M_1^+, 1, 2, M_2^+ is a candidate for a set of points reachable from **I** (reachable set).

(ii) Region M_4^-, 4, 3, M_3^- is a candidate for a set of points controllable to **I** (controllable set).

(iii) Region 1, A, 2, 1 is a candidate for a set of points which collide with **I** (collision set).

(iv) Region 3, B, 4, 3 is a candidate for a set of points shadowed by **I** (shadow set).

(v) Regions 1, C, 4, 1 and 2, D, 3, 4 are candidates for a set of points ambivalent with respect to **I** (Ambivalent set).

(vi) etc.

The boundary of the collision set represents a dividing line between a situation where there is no real risk of collision with a target standing on, and a clear risk of collision with a target standing on. A target standing on located outside the collision set can be kept outside the SPC by proper manoeuvering by own ship. A target standing on located inside the collision set can not be kept from penetrating the SPC by own ship manoeuvres.

On way of locating the collision set boundaries is to run system trajectories along it. Since H is maximized along 1A and 2A, the control program, by own ship, which will run system trajectories along it may be found by employing necessary conditions for control on the boundary of the reachable set to SPC. However instead of integrating these conditions forward in time, retro integration must be used.

MACRO COLLISION AVOIDANCE

The determination of the boundaries of the collision set is simpler if the system dynamics are written in terms of a rotating coordinate system. Equations (1) and (2) are written in terms of a stabilized radar reference frame with respect to own ship. If instead, the motion of a target blip are written in terms of a non-stabilized radar reference frame (rotating) with respect to own ship, then equations (1) and (2) are replaced by

$$\dot{x}_1 = s_t \sin x_3 - v x_2 \qquad (18)$$

$$\dot{x}_2 = s_t \cos x_3 - s_0 + v x_1 \qquad (19)$$

where $x_3 = \gamma_t - \gamma_0$. Thus equations (3) and (4) are given by the single expression

$$\dot{x}_3 = u - v . \qquad (20)$$

For convenience, let $\gamma_0 = 0$. The macro collision avoidance assumption of target standing on is introduced by setting $u = 0$. Let

$$H = \lambda_1 (s_t \sin x_3) + \lambda_2 (s_t \cos x_3 - s_0) + \sigma_v v \qquad (21)$$

$$\sigma_v = \lambda_2 x_1 - \lambda_1 x_2 - \lambda_3 \qquad (22)$$

then necessary conditions for determining a control which will drive the system along the boundary of the reachable set from an SPC of radius R defined by

$$x_1^2 + x_2^2 = R^2 \qquad (23)$$

are given by (18), (19), and

$$\dot{\lambda}_1 = -\lambda_2 v \tag{24}$$
$$\dot{\lambda}_2 = \lambda_1 v \tag{25}$$
$$\dot{\lambda}_3 = s_t(\lambda_2 \sin x_3 - \lambda_1 \cos x_3) \tag{26}$$

$$v = \begin{cases} v_{max} & \text{if } \sigma_v > 0 \\ 0 & \text{if } \sigma_v = 0 \text{ over a nonzero time interval} \\ -v_{max} & \text{if } \sigma_v < 0 \end{cases} \tag{27}$$

$$H = 0 \tag{28}$$
$$\lambda_1(0) = 2\mu x_1(0) \tag{29}$$
$$\lambda_2(0) = 2\mu x_2(0) \tag{30}$$
$$\lambda_3(0) = 0 \tag{31}$$

where μ is a positive constant [Grantham and Vincent (1975)]. These conditions when integrated backward in time are then necessary for a control which will drive the system along the boundary of the collision set.

The singular solution $v = 0$ in (27) was obtained from the second time derivative of the switching function

$$\dot{\sigma}_v = s_0 \lambda_1 \tag{31}$$
$$\ddot{\sigma}_v = -s_0 \lambda_2 v \tag{32}$$

Setting $\sigma_v = 0$ it follows that either $\lambda_2 = 0$ or $v = 0$. Since $\lambda_2 = 0 \Rightarrow \lambda_1 = 0$ from (21) and thus $\lambda_3 = 0$ from (22), it follows that the singular solution is given by $v = 0$. Substituting $\sigma_v = \sigma_v = -s_0 \lambda_1 = 0$ in (28) yields the result that

$$\lambda_2 (s_t \cos x_3 - s_0) = 0 \tag{33}$$

along a singular arc. Again $\lambda_2 \neq 0$ so that $\cos x_3 = s_0/s_t$. Thus if $s_0 > s_t$ there will be no singular solution. The further necessary condition of Kelley *et al.* (1967) is used for the case where $s_0 < s_t$

$$\frac{\partial}{\partial v} (\ddot{\sigma}_v) = -s_0 \lambda_2 \geq 0. \tag{34}$$

It follows from (30) that terminal singular control is possible if $x_2(0) < 0$.

Since there are three state coordinates, the boundary of the collision set is a two dimensional surface in x_1, x_2, x_3 space. In order to obtain useful information with respect to the radar display, cross-sections of the boundary of the collision set may be projected on to the $x_1 - x_2$ plane for constant values of x_3. A qualitative guide to the solutions may now be obtained by examining control possibilities on the SPC. Substituting (29) and (30) into (28) yields

$$x_2(0)/x_1(0) = s_t \sin x_3(0)/(s_0 - s_t \cos x_3(0)) . \tag{35}$$

Since $\sigma_v(0) = 0$, control on the SPC must be obtained from

$$v(0) = \begin{cases} v_{max} & \text{if } \dot{\sigma}_v(0) = s_0\lambda_1(0) < 0 \\ 0 & \text{if } \dot{\sigma}_v(0) = s_0\lambda_1(0) = 0 \quad \text{over a nonzero time interval} \\ -v_{max} & \text{if } \dot{\sigma}_v(0) = s_0\lambda_1(0) > 0 \end{cases} \qquad (36)$$

By letting $x_3(0)$ range over $0 - 2\pi$, the points where the collision boundary is tangent to the SPC are obtained. This is illustrated in figure 3 for both $s_0 > s_t$ and $s_0 < s_t$

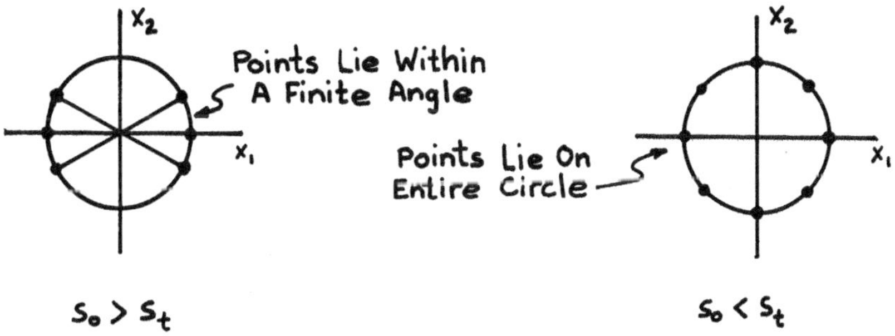

FIGURE 3. COLLISION BOUNDARY POINTS ON THE SPC

When $s_0 > s_t$ there is a symmetry to the solutions starting on the left side of the SPC to those starting from the right side of the SPC. There is no singular solution. On the right side $v(0) = -v_{max}$ (see equations (29) and (36)) and on the left side $v(0) = v_{max}$. Thus a projection of right side collision boundary trajectories on to $x_2 - x_3$ in the neighborhood of the SPC would be as indicated in figure 4.

When $s_0 < s_t$ the two solutions are again symmetric. Solutions in the neighborhood of SPC on the right boundary (w.r.t. intersection) are as in Figure 4. The singular solution occurs when (33) is satisfied. Let

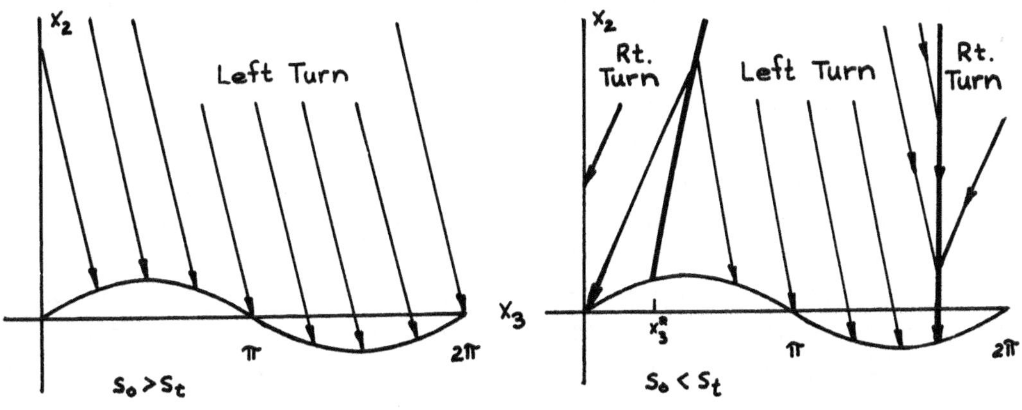

FIGURE 4. TRAJECTORIES ON THE RIGHT COLLISION BOUNDARY

$x_3^* = \arcos\ (s_0/s_t)$. It follows from (29), (35), and (36) that on the right side of the SPC $v(0) = v_{max}$ for $0 \leq x_3 < x_3^*$ and $2\pi - x_3^* < x_3 \leq 2\pi$ and $v_0 = -v_{max}$ for $x_3^* < x_3 < 2\pi - x_3^*$. Equation (34) is satisfied for $2\pi - x_3^*$ but not for x_3^*. Thus a singular arc is located at $2\pi - x_3^*$ and a dispersal arc is located at x_3^*. The solutions are reversed for the left side. The dispersal arc is at $2\pi - x_3^*$ and the singular arc is at x_3^*.

In both cases, the collision boundaries intersect (as in figure 2) before the structure of the solution changes from that indicated in figure 4. Thus if $s_0 > s_t$, boundary control is either a hard left turn on a hard right turn. A left turn is made should a target reach the right boundary and a right turn is made should a target reach the left boundary. These turns are maintained until the target reaches the SPC. If $s_0 < s_t$, boundary control will initially be a hard right or left turn, however, now because of the singular solution, the turn may be terminated (followed by $v = 0$) before the target reaches the SPC. Although trajectories which terminate on the right side of the SPC will do so with a hard left turn and trajectories which terminate on the left side of the SPC will do so with a hard right turn, these turns can now no longer be associated with a right or left "side" of the collision boundary.

Miloh (1974) has obtained cross sections for speed rations of $s_t/s_0 = 1/\sqrt{2}$ and $s_t/s_0 = \sqrt{2}$. His results are typical for the two cases $s_0 > s_t$ and $s_0 < s_t$. A qualitative sketch of these two cases are illustrated in figure 5. The results in this form are quite useful as the actual manoeuvres to keep the system on the boundary of the collision set can also be indicated. Miloh (1974) gives detailed information in this regard.

The union of all the collision boundary cross sections, UBC, may be superimposed on the stabilized radar display as indicated in figure 6. The positive x_2 axis used to obtain the cross sections is now aligned with the heading flasher. Only two UBC curves are shown. The assumption is that all targets slower than own ship are at one speed s_{ts} and all targets faster than own ship are at another speed s_{tf} The two lines drawn at an angle $\alpha = \arcsin\ (s_{ts}/s_0)$ with respect to the heading flasher define a cone within which a slower target can be on a collision course. Any target directed toward the SPC outside this cone must be faster than own ship. Thus target A could be faster at slower than own ship whereas target B must be faster than own ship.

By combining the result of figure 5 with the display as in

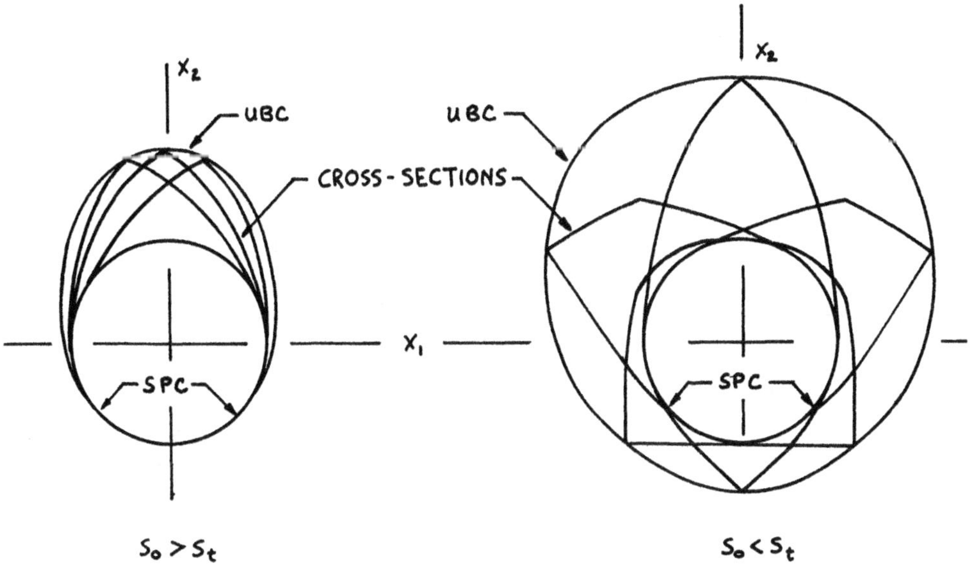

FIGURE 5. CROSS-SECTIONS OF COLLISION BOUNDARY

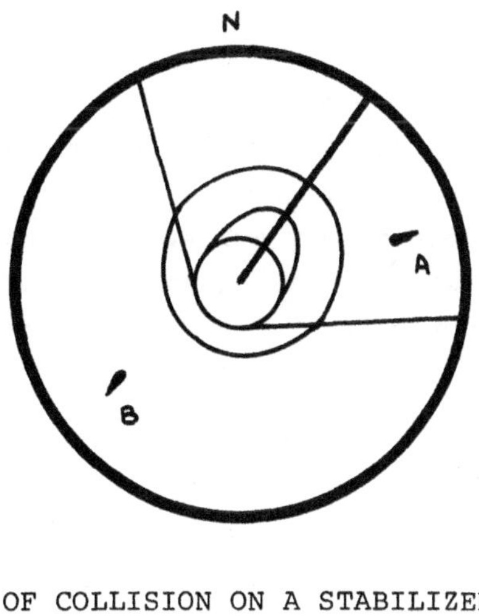

FIGURE 6. RISK OF COLLISION ON A STABILIZED RADAR DISPLAY

figure 6, a procedure for collision avoidance is suggested. Suppose that a number of targets enter the radar screen with one or more directed toward the SPC. At first the pilot could use his collision avoidance training and experience to make judgements as to the proper course of action. However, should a target, faster than own ship, approach the outer UBC or a target, slower than own ship, approach the inner UBC, he would then know that this may be the last opportunity for him to manoeuvre in such a way so as to keep the target out of the SPC. Given the relative heading of the target, he could call for the appropriate collision boundary cross-section with its corresponding turning schedule as a guide for what control action to take. Even then, it may be that the suggested manoeuvres would not be successful in keeping the target out of the SPC (the target may also be manoeuvering) or some other target may be driven into the SPC because of the turn. Should either of these situations occur, the risk of collision with the target inside the SPC would then be sufficiently high as to require micro collision avoidance information.

Micro Collision Avoidance

To make progress, the optimistic assumption is made that only one target enters the SPC. However by virtue of the fact that a target has penetrated the SPC it is now wise to take the pessimistic view that the pilot of the target ship is either confused with, or unaware of the present situation and may operate his ship in such a way as to cause a collision.

Depending upon the relative heading and speed of the target, the SPC may be divided into points from which the target can collide with own ship, despite any manoeuvre by own ship and points from which own ship can avoid a collision, despite any manoeuvre by the target. The dividing line between these two cases if of interest as it can be used as a guide for identifying appropriate manoeuvres. The dividing line is a barrier for a game of kind [Isaacs (1965)].

Again the barrier may be determined if controls can be found which will run system trajectories along it. However in this case the system trajectories will end on a surface which simulates the actual geometric configuration of the ships at contact, which is own ships shape appropriately displaced to account for the physical dimension of the target ship. The resultant surface is not necessarily smooth and may require special boundary conditions as given by Peng (1973).

The equations for the target ship are again given by (18)-(20). Let

$$H = \lambda_1 s_t \sin x_3 + \lambda_2 (s_t \cos x_3 - s_0) + \sigma_u u + \sigma_v v \tag{37}$$

$$\sigma_u = \lambda_3 \tag{38}$$

$$\sigma_v = \lambda_2 x_1 - \lambda_1 x_2 - \lambda_3 , \tag{39}$$

then necessary conditions for determining controls u and v which will drive the system along the barrier [Blaguière, *et al.* (1969)] are given by (24)-(28) and

$$u = \begin{cases} u_{max} & \text{if } \sigma_u < 0 \\ 0 & \text{if } \sigma_u = 0 \text{ over a nonzero time interval} \\ -u_{max} & \text{if } \sigma_u > 0 \end{cases} \tag{40}$$

plus appropriate boundary conditions.

The singular control $u = 0$ in (40) was obtained from the second time derivative of the σ_u switching function

$$\dot{\sigma}_u = s_t(\lambda_2 \sin x_3 - \lambda_1 \sin x_3) \tag{41}$$

$$\ddot{\sigma}_u = s_t u(\lambda_1 \sin x_3 + \lambda_2 \cos x_3) . \tag{42}$$

Setting $\ddot{\sigma}_u = 0$ it follows that either $\lambda_1 \sin x_3 + \lambda_2 \cos x_3 = 0$ or $u = 0$. The first choice plus $\dot{\sigma}_u = 0$ implies $\lambda_1 = \lambda_2 = 0$ and $\lambda_3 = 0$ from $\dot{\sigma}_u = 0$. Thus the singular solution is given by $u = 0$. Requiring that $\sigma_u = \dot{\sigma}_u = 0$ in (28) yields

$$(s_t - s_0 \cos x_3) = v(x_2 \sin x_3 - x_1 \cos x_3) . \tag{43}$$

A further necessary condition on singular control $u = 0$ is given by

$$\frac{\partial}{\partial u}(\ddot{\sigma}_u) = s_t(\lambda_1 \sin x_3 + \lambda_2 \cos x_3) \leq 0 . \tag{44}$$

It appears that singular control will be part of the barrier control program for the target ship for both $s_0 > s_t$ and $s_0 < s_t$.

Vincent (1975) obtained barrier cross sections using the above necessary conditions with trajectories terminating on a "ship shape" and with $s_0 > s_t$. A qualitative sketch of the cross sections and the projection of the trajectories onto $x_2 - x_3$ are illustrated in figure 7. It is assumed that the radius of the SPC shown is sufficiently large so that the left and right barriers intersect inside the SPC. As with the micro collision avoidance case of $s_0 > s_t$, own ship makes a hard left turn or hard right turn depending upon whether the target is on the 'right" or "left" barrier. However in this case the turn is not necessarily maintained up to the point of contact. If the trajectory terminates on the rear part of own ship then a last minute switch in control is required. For example, figure 7 illustrates that for $\pi < x_3(t_f) < 2\pi$ a hard left turn is used by own ship until near contact

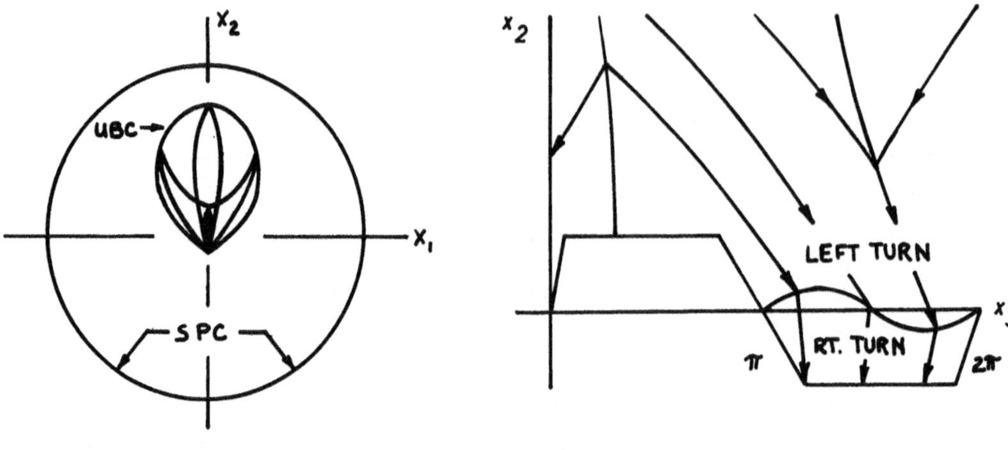

BARRIER CROSS-SECTIONS RIGHT BARRIER

FIGURE 7. MICRO COLLISION AVOIDANCE $s_0 > s_t$

then control by own ship is switched to a hard right turn. The effect is to swing the vulnerable rear portion of the ship away from the target.

Cross-sections for the case where $s_0 < s_t$ have not yet been obtained. In this case the target would be less manoeuvrable, i.e. have a larger turning radius than own ship.

COMMENT

It is suggested that collision avoidance procedures can be employed at three different levels. At the first level, the control of own ship is left to the skill of the pilot who must judge the risk of collision in terms of the rules of the road, multiple targets, and any additional physical constraints imposed on his choice of control action. He would attempt to maintain targets outside an appropriate UBC. Should a target reach a UBC, the second procedural level begins. Location, speed, and heading information about the target is used to calculate a collision surface cross-section. This cross-section is then projected onto the stabilized radar display along with a suggested avoidance manoeuvre. The pilot may then act accordingly. The third procedural level is implemented if a target enters the SPC. Again location, speed, and heading information is now used to calculate a barrier cross-section. In this case hope for saving the ships from collision may depend on precise utilization of the micro collision avoidance information as displayed on the stabilized radar screen. In this sense micro collision avoidance is used to obtain backup information when macro collision avoidance fails.

REFERENCES

Blaquiere, A., Gerard, F. and Leitmann, G., *Quantitative and Qualitative Games*, Academic Press, New York, 1969.

Carpenter, M.H. and Waldo, W.M., *Automated Collision Avoidance - A New Look at an Old Problem*, Maritime Institute of Technology and Graduate Studies, Linthicum Heights, Maryland, 1974.

Grantham, W.J. and Vincent, T.L., "A Controllability Minimum Principle", *Journal of Optimization Theory and Applications*, Vol. 17, Nos 1/2, pp. 93-114, October, 1975.

Isaacs, R., *Differential Games*, John Wiley and Sons, Inc. New York, 1965.

Kelley, H.J., Kopp, R.E., Moyer, G.H., "Singular Extremals, *Topics in Optimization*, Edited by G. Leitmann, Academic Press, New York, 1967.

Miloh, T., *Determination of Critical Manoeuvres for Collision Avoidance Using the Theory of Differential Games*, Institut für Schiffbau Der Universität Hamburg, Bericht Nr. 219, November 1974.

Miloh, T. and Sharma, S.D., *Maritime Collision Avoidance as a Differential Game*, Instut für Schiffbau Der Universität Hamburg, Bericht Nr. 329, July 1975.

Peng, W.Y., *Controllability and Qualitative Game Transversality Conditions for Non-Smooth Targets*, Ph.D. Dissertation, University of Arizona, Tucson 1973.

Vincent, T.L., Cliff, E.M., Grantham, W.J., and Peng W.Y., *A Problem of Collision Avoidance*, University of Arizona, Tucson, EES Series Report No. 39. November 1972.

Vincent, T.L. and Peng, W.Y., *Ship Collision Avoidance*, Navy Workshop in Differential Games, Annapolis Maryland, 1973.

Vincent, T.L., "Avoidance of Guided Projectiles", *The Theory and Application of Differential Games*, J.D. Grote (ed.) D. Reidel, Dordrecht, pp. 267-279, 1975.

Webster, W.C., *When is a Collision Unavoidable?* Tenth Naval Hydrodynamics Symposium, Boston, Mass. 1974.

ENVIRONMENTAL ADAPTATION BY ANNUAL PLANTS

(An optimal control/games viewpoint)

Thomas L. Vincent

Aerospace and Mechanical Engineering
University of Arizona, Tucson 85721/USA

ABSTRACT. As with all living things, annual plants have different ways of adapting to the environment. Adaptation may occur via: (1) physiological versatility, (2) developmental flexibility and (3) genetic evolution. Adaptation (1) is associated with a rapid response to short term changes in the environment, such as the opening or closing of the stomata in the leaves of plants with a change in the ambient temperature or humidity. Adaptation (2) is associated with response due to a more long term conditioning by the environment, such as limited flowering of a plant due to a dry season. Adaptation (3) is the genetic adaptation in response to particular semi-permanent conditions in the environment. If these conditions should change, then it may take several seasons to produce a new genetic adaptation.

Adaptations (2) and (3) are examined here in the context of two different habitats, that of growth under limited nutrient and growth under limited sunlight. The viewpoint is taken that survival is a direct consequence of maximizing seed production.

In the limited nutrient habitat, it is assumed that a nutrient conversion efficiency represents a control parameter subject to genetic selection. In the limited sunlight habitat, it is assumed that the percent allocation of total growth into seeds is a control variable subject to developmental flexibility. In each case, controls which result in maximum seed production are determined. Results are compared between a monoculture and a mixture of two different species.

INTRODUCTION

Smith (1975) adopts the terms, physiological versatility, developmental flexibility, and genetic adaptability to describe three kinds of biological adaptation. Roughly speaking, these terms may be thought of as the grouping of a continuum of response into short term, mid term, and long term responses. The notion is convenient from a controls point of view as each type of adaptation may then be identified with a particular type of control variable.

Under the notion that physiological versatility represents a short-term response to changes in the environment, it follows that implementation of such a response would require that the biological system be able to sense not only its own current state, but the current state of the environment as well. Since longer term (or seasonal) changes in the environment would be reflected in the state of the biological system, a response via developmental flexibility would require only the sensing of the biological system's own state. Genetic evolution (adaptation) represents a response to changes in semi-permanent conditions of the environment which have already been adapted for. In this case, no biological sensing mechanism is required. Adaptation results from the concentration of favourable genes already existing in the biological system, or more rarely, through mutation [Pimentel D, (1968)]. For annual plants the concentration process is easily visualized in terms of the greater seed productivity of plants with favourable genes.

These concepts are easily summarized in mathematical terms. Let y represent the environmental state and x represent the biological state at time t. Let $v(\cdot)$ represent control available through physiological versatility, $u(\cdot)$ represent control available through developmental flexibility, and β represent control resulting from genetic evolution. In these terms the dynamics of the biological system is envisioned to be given by

$$dx/dt = f(x,y,\ v(x,y,t),\ u(x,t),\ \beta) \tag{1}$$

where $f(\cdot)$ is an appropriate function of the variables indicated.

Given a mathematical model such as (1), one may then determine how to implement the controls in order to satisfy a hypothesised criteria. Adaptation is taken to be the response of the controls under a specified criteria to environmental inputs.

Genetic evolution (β) and developmental flexibility $[u(\cdot)]$ will be examined separately in two idealized plant-environmental situations. A model for each situation is developed from the hypothesis that plant growth results from the conversion of resource material into plant material. Let

p = biomass of plant per unit area

A = amount of resources taken in per unit plant biomass per unit time

β = conversion factor (a measure of "conversion efficiency")

α/β = resource maintenance requirement due to respiration

then one way of expressing the above hypothesis is to write

$$dp/dt = (\beta A - \alpha)p \; . \tag{2}$$

Each model will contain the assumption that there is but one resource which is limiting with respect to the growth of the plant. For the sake of discussion, the resource in the first model is thought of as the amount of nutrient in the soil and the resource in the second model is thought of as the amount of sunlight available.

A FORAGED RESOURCE MODEL

The roots of a plant take up nutrients from the soil as required by the photosynthesis process. The root may be thought of as foraging for the nutrient. Thus the functional relationship between nutrient uptake and the amount of nutrient available in the soil would be similar to functions used in any foraged resource situation [De Angelis, *et al* (1975)]. When the nutrient is abundant the uptake of nutrient is assumed to be at some maximum rate. This saturation effect is accounted for by writing

$$A/A_m = e_n N/(1 + e_n N) \tag{3}$$

A_m = Maximum value for A
e_n = foraging factor (a measure of "foraging efficiency")
N = amount of nutrient in soil.

Data is available [Trenbath (1976)] which indicates that there is virtually no maintenance requirement for nutrient. Although less conclusive, the experimental data also suggests a trade off between e_n and β of the form $e_n \sim 1/\beta^2$. For the present, these observations will be cannonized by assuming

$$\alpha = 0 \tag{4}$$

$$A_m e_n = \gamma/\beta^2 \tag{5}$$

where γ is a constant associated with a particular species. Thus equation (2) becomes

$$dp/dt = \gamma N p/\beta(1 + e_n N) \; . \tag{6}$$

The rate of change of nutrient in the soil is given by

$$dN/dt = \text{input} - Ap \; . \tag{7}$$

Attention will be confined to the limited nutrient case defined here by the conditions that $e_n N \ll 1$ and that the input to equation (7) is zero. It then follows that equation (7) may be integrated to yield

$$N + p/\beta = \ell \tag{8}$$

where ℓ is a constant set by the initial conditions. Equation (6) reduces to

$$dp/dt = \gamma(\ell - p/\beta)p/\beta \,. \tag{9}$$

Equation (9) may be alternately written as

$$dp/dt = r(k - x)x/k \tag{10}$$

where $r = \gamma\ell/\beta$ and $k = \ell\beta$. This is the familiar logistics equation [May (1974)] except that here the carrying capacity k and the intrinsic growth rate r are identified in terms of particular functions of the conversion factor β. Indeed as β changes, so does the trade off between r and k. Pianka (1974) gives biological arguments for the existence of such a trade off.

The model may be extended to two different species of plants using the same limited nutrient by writing (7) as

$$dN/dt = -A_1p_1 - A_2p_2 \tag{11}$$

where the subscripts refer to the individual species. Thus the dynamical equations become

$$dp_1/dt = \gamma_1(\ell - p_1/\beta_1 - p_2/\beta_2)p_1/\beta_1 \tag{12}$$

$$dp_1/dt = \gamma_2(\ell - p_1/\beta_1 - p_2/\beta_2)p_2/\beta_1 \tag{13}$$

By letting $r_1 = \ell\gamma_1/\beta_1$, $k_1 = \ell\beta_1$, $a_{12} = \beta_1/\beta_2$, $r_2 = \ell\gamma_2/\beta_2$, $k_2 = \ell\beta_2$, $a_{21} = \beta_2/\beta_1$ these equations may be written in form of the Lotka-Volterra competition equations

$$dp_1/dt = r_1(k_1 - x_1 - a_{12}x_2)x_1/k_1 \tag{14}$$

$$dp_1/dt = r_2(k_2 - a_{21}x_1 - x_2)x_2/k_2 \tag{15}$$

see for example Pianka (1974).

In what follows it is more convenient to use equations (12) and (13) in a nondimensional form. Let $x_1 = p_1/\ell$, $x_2 = p_2/\ell$, $\tau = \gamma_1 t$, and $c = \gamma_2/\gamma_1$, then

$$dx_1/d\tau = x_1(1 - x_1/\beta_1 - x_2/\beta_2)/\beta_1 \tag{16}$$

$$dx_2/d\tau = cx_2(1 - x_1/\beta_1 - x_2/\beta_2)/\beta_2 \tag{17}$$

A NONFORAGED RESOURCE MODEL

The leaves of a plant intercept sunlight for the photosynthesis process. If the canopy of a plant is sufficiently dense, then all of the incoming light will be intercepted. This saturation effect will be accounted for by writing

$$Ap/I = e_I p/(1 + e_I p) \tag{18}$$

I = light available per unit area per unit time

e_I = light gathering factor (a measure of "gathering efficiency").

Since it is the leaves which gather the light, it is implied by (18) that the plant is mostly leaf. Substituting (18) into (2) yields

$$dp/dt = (\beta\, e_I\, I/(1 + e_I p) - \alpha)p \tag{19}$$

If two different species of plants are growing together (using the same sunlight resource) then one way to express the effect of one plant on the other is to write

$$dp_1/dt = (c_1/(1 + c_5 p_1 + c_6 p_2) - c_3)p_1$$
$$dp_2/dt = (c_2/(1 + c_5 p_1 + c_6 p_2) - c_4)p_2 \tag{21}$$

where $c_1 = \beta_1 e_{I_1} I$, $c_2 = \beta_2 e_{I_2} I$, $c_3 = \alpha_1$, $c_4 = \alpha_2$, $c_5 = e_{I_1}$, and $c_6 = e_{I_2}$. Except for the constants c, the numbered subscripts refer to the individual species.

GENETIC EVOLUTION IN A MONOCULTURE

By setting $x_2 = 0$ in equation (16) the growth of a single species of plant under a limited nutrient environment is obtained. For convenience the subscript is dropped,

$$dx/d\tau = (1 - x/\beta)x/\beta \tag{22}$$

Note from equation (8) that $p/\beta \to \ell$ corresponds to $N \to 0$ and that if the initial value of p is small then ℓ is approximately equal to the initial nutrient in the soil. Thus equation (22) fits the notion of an annual plant by simply employing appropriate boundary conditions. Equation (22) is assumed to be valid over the growing season. At the end of the season the nutrient in the soil is readjusted to its starting value by resetting x to its initial value.

Assume now that the criteria for survival is that seed production be maximized. If seed production is proportional to biomass at the end of the growing season, then the appropriate value for β is one which maximizes the integral

$$\int_0^T (dx/d\tau)\, d\tau \tag{23}$$

where T is the length of the growing season. Let

$$H = (1 + \lambda)[(x/\beta) - (x/\beta)^2] \tag{24}$$

then necessary conditions for determining the value of β to maximize

(23) as obtained from the maximum principle (under parameter control) [see Leitmann (1966)] are

$$\partial H/\partial x = (1 + \lambda)[1 - 2(x/\beta)]/\beta = -d\lambda/d\tau \tag{25}$$

$$\int_0^T (\partial H/\partial\beta)d\tau = - \int_0^T (1 + \lambda)[(x/\beta) - 2(x/\beta)^2]/\beta \, d\tau = 0 \tag{26}$$

$$H = (1 + \lambda)dx/d\tau = k \quad (k = \text{constant}). \tag{27}$$

The above relations are to be satisfied subject to boundary conditions. Two cases will be examined. One can take the view that at the beginning of each season, a given number of "sites" are available which are favourable for the germination and growth of a seed. The genotype which produces the most seeds will have the greatest probability of filling the majority of sites. Thus genetic evolution will still operate under the criteria of maximum seed production even though the number of sites are fixed and all the sites are filled. A second view is that a given fraction of seeds produced at the end of one season survive to germinate at the beginning of the next season.

In the first case, the boundary condition on the initial value of the state and the final value of the adjoint variable are given by

$$x(0) = \text{constant}\,, \quad \lambda(T) = 0 \tag{28}$$

From (25), (26), and (28) there is obtained

$$\int_0^T x(d\lambda/d\tau)d\tau = -[x(0)\lambda(0) + \int_0^T \lambda(dx/d\tau)d\tau] = 0 \tag{29}$$

Integrating (27) over $[0,T]$ using (28) and (29) yields

$$x(T) - x(0)[1 + \lambda(0)] = kT\,. \tag{30}$$

Use of (30) with (27) evaluated at the initial and final times yields a relationship between $x(0)$, $\lambda(0)$, and β. Thus given T and $x(0)$, one can guess a value for β, calculate $\lambda(0)$ and then integrate equations (22) and (15) forward in time checking to see if (26) is satisfied at the final time. This procedure will ultimately yield β. However, it is simpler to integrate (22) to obtain

$$x(T) = \beta \exp(T/\beta)/[\beta/x(0) - 1 + \exp(T/\beta)] \tag{31}$$

Differentiating with respect to β and setting the result equal to zero gives

$$\exp(T/\beta) = 1 + (\beta/x(0) - 1)T/\beta\,. \tag{32}$$

Equations (31) and (32) were used to make the calculations of table 1. The ratio $T/x(0)$ is the quantity given by the problem statement. It represents the environmental conditions or habitat in which the plant finds itself.

$T/x(0)$	8.34	20.65	57.60	152	2988
T/β	2.00	2.93	4.00	5.00	8.00
$x(T)/\beta$	.6982	.7558	.8029	.8342	.8889

Table 1 Fixed number of sites

It follows from the above calculations that

(i) Given $x(0)$, increasing T increases $x(T)$ and β

(ii) Given T, increasing $x(0)$, increases $x(T)$ and β.

Thus a plant will tend for k selection (large β) with a longer growing season and/or larger number of sites. Note over this wide range of values for $T/x(0)$, the ratio $x(T)/\beta$ has a small range of values. Having $x(T)/\beta$ equal to unity corresponds to all the nutrient utilized.

The alternative boundary condition: $x(0) = \mu x(T)$ where $0 < \mu < 1$ is mixed. The corresponding necessary condition on the adjoint variable [Vincent and Mason (1969)] is given by

$$\lambda(T) = \mu\lambda(0) . \tag{33}$$

From (25), (26) and (33) it follows that

$$\int_0^T x(d\lambda/d\tau)d\tau = -\int_0^T \lambda(dx/d\tau)d\tau = 0 . \tag{34}$$

Integrating (27) over $[0,T]$ using (33) and (34) yields

$$x(T)(1 - \mu) = kT . \tag{35}$$

Assuming that $\mu \ll 1$, k is approximated by $k = x(T)/T$. Use of (35) with (22) evaluated at the initial and final times, utilizing $\mu \ll 1$, yields

$$x(T)/\beta = (T/\beta)/(1 + T/\beta) . \tag{36}$$

This result used with (27) evaluated at the final time yields

$$\lambda(T) = (\beta/x(T)) - 1 . \tag{37}$$

Given T, one can guess β, calculate $x(T)$ and $\lambda(T)$ from (36) and (37) and then integrate equations (22) and (25) backward in time, checking to see if (26) is satisfied at the initial time. Again it is simpler to find β directly by substituting $x(0) = \mu x(T)$ into (31), solving for $x(T)$, differentiating with respect to β and setting the result equal to zero. One obtains

$$\exp(T/\beta)[\mu \exp(T/\beta) - (1+\mu) + (T/\beta)(\mu - 1)] + 1 = 0 . \tag{38}$$

Under the assumption that $\mu \ll 1$ this reduces to

$$\mu \exp(T/\beta) = 1 + T/\beta . \tag{39}$$

This latter expression slightly overestimates the value of T/β for a given $\mu = x(0)/x(T)$. Equations (36) and (39) were used to make the calculations of table 2.

μ	.406	.1031	.092	.040	.003
T/β	2.00	2.93	4.00	5.00	8.00
$x(T)/\beta$	.667	.7455	.8000	.8333	.8889

Table 2 Fraction of seeds survive

It is observed that

(i) Given μ , increasing T increases $x(T)$ and β .

(ii) Given T , increasing μ increases $x(T)$ and β .

Given $\mu = x(0)/x(T)$ and T both β and $x(0)$ are determined via table 2. One may then enter table 1 with this $T/x(0)$ and observe that the same value for β is obtained. In other words, the optimum value for β for $x(0) = \mu x(T)$ boundary condition is also optimal for the $x(0)$ = constant boundary condition, provided that T and $x(0)$ are the same. The two tables are equivalent.

Since nondimensional time $\tau = \gamma t$, it follows that if T^* is the growing season in days then $T = \gamma T^*$. Thus in comparing two plants with different values for γ , both with the same growing season T^*, the one with the larger γ will have a larger value for T and consequently β . The faster growing plant (larger γ) will have the larger conversion factor β .

The solutions obtained here under the periodic boundary conditions are stable with respect to variations in $x(0)$. If T , β , and μ are kept fixed, the solution to (22) will return to the optimal value of $x(0)$ after a few seasons. Such stability is necessary for the process of genetic evolution to take place.

GENETIC EVOLUTION IN MIXTURE

Consider two different species of plants growing in mixture using the same limited resource whose dynamics over the growing season are given by equations (16) and (17). Assume that each species is operating under the criteria for survival that its own seed production be maximized. Assume again that seed production is proportional to biomass at the end of the growing season. Ruling out cooperation between the species, it is hypothesised that genetic evolution will yield values for β_1 and β_2 such that the final biomass of the first species is maximized with respect to variations in β_1 and the final biomass of the second species is maximized with respect to β_2. This is the Nash equilibrium game concept [Nash 1951)].

Leitmann (1974) gives necessary conditions for a Nash solution for dynamical systems. These conditions may be used if β_1 and β_2 are interpreted to be state variables with state equations $\dot{\beta}_1 = B_1 = 0$ and $\dot{\beta}_1 = B_2 = 0$. Let

$$G_1 = x_1(1 - x_1/\beta_1 - x_2/\beta_2)/\beta_1 \tag{40}$$

$$G_2 = cx_2(1 - x_1/\beta_1 - x_2/\beta_2)/\beta_2 \tag{41}$$

$$H^1 = (1 + \lambda_3)G_1 + \lambda_4 G_2 + \lambda_5 B_1 \tag{42}$$

$$H^2 = \lambda_6 G_1 + (1 + \lambda_7)G_2 + \lambda_8 B_2 \tag{43}$$

then necessary conditions for determining values of β_1 and β_2 for a Nash solution are

$$\partial H^1/\partial x_1 = (1+\lambda_3)\partial G_1/\partial x_1 + \lambda_4 \partial G_2/\partial x_2 = -d\lambda_3/d\tau \tag{44}$$

$$\partial H^1/\partial x_2 = (1+\lambda_3)\partial G_1/\partial x_2 + \lambda_4 \partial G_2/\partial x_2 = -d\lambda_4/d\tau \tag{45}$$

$$\partial H^1/\partial \beta_1 = (1+\lambda_3)\partial G_1/\partial \beta_1 + \lambda_4 \partial G_2/\partial \beta_1 = -d\lambda_5/d\tau \tag{46}$$

$$H^1 = k_1 \tag{47}$$

$$\partial H^2/\partial x_1 = \lambda_6 \, \partial G_1/\partial x_1 + (1+\lambda_7)\partial G_2/\partial x_1 = -d\lambda_6/d\tau \tag{48}$$

$$\partial H^2/\partial x_2 = \lambda_6 \, \partial G_1/\partial x_2 + (1+\lambda_7)\partial G_2/\partial x_2 = -d\lambda_7/d\tau \tag{49}$$

$$\partial H^2/\partial \beta_2 = \lambda_6 \, \partial G_1/\partial \beta_1 + (1+\lambda_7)\partial G_2/\partial \beta_1 = -d\lambda_8/d\tau \tag{50}$$

$$H^2 = k_2 \tag{51}$$

$$\lambda_5(0) = \lambda_8(0) = 0 \tag{52}$$

$$\lambda_5(T) = \lambda_8(T) = 0 \tag{53}$$

where k_1 and k_2 are constants. Boundary conditions on the "real" state variables x_1 and x_2 must now be specified. As in the monoculture case, one can think in terms of fixed number of sites for each species or in terms of a given seed survival rate for each species. Other combinations are also possible, e.g. one could fix the total number of sites and allow them to be filled by either species. Only the first case will be considered here. In this case the additional boundary conditions are given by

$$x_1(0) = \text{constant}, \quad x_2(0) = \text{constant} \tag{54}$$

$$\lambda_3(T) = \lambda_6(T) = 0 \tag{55}$$

$$\lambda_4(T) = \lambda_7(T) = 0\,. \tag{56}$$

A computer solution requires guessing values for β_1, β_2, $x_1(T)$, and $x_2(T)$, then integrating (16), (17), (44) - (46), and (48) - (50) backward in time checking to see if at the initial time (52) is satisfied along with the initial conditions $x_1(0)$ and $x_2(0)$.

Consider two fields, each which receives ℓ units of nutrient at the beginning of each season. Consider now two species with $\gamma_1 = \gamma_2 = .0236$. Suppose that there are the same number of sites for each species in each field: $x_1(0) = x_2(0) = .2057$. Let the growing season be 180 days. Suppose that in the first field species 1 is growing in monoculture and in the second field species 2 is growing in monoculture. Thus for each field $T = 180(.0236) = 4.25$ and

$$T/x_1(0) = T/x_2(0) = 20.66 , \tag{57}$$

Assuming that each species has adapted to its monoculture environment, the yield and β factor for each species is found in table 1. Suppose now that each species "invades" the other field and that each species adapts to this new situation according to the Nash solution. A Nash solution is obtained with $\beta_1 = \beta_2 = 1$, $x_1(T) = x_2(T) = .49$. The results are summarized in table 3. The subscripts have been dropped. The ratios refer to either species.

	T/β	$x(T)/\beta$	$x(T)/T$
Monoculture	2.93	.7558	.2579
Mixture	4.25	.4900	.1153

Table 3 A comparison between monoculture and mixture

The conversion efficiency β decreases in mixture, resulting in faster growth rates (a move toward r selection). As would be expected, the individual yields in mixture are less than in monoculture. The total yield in mixture

$$2\ x(T)/T = .2306 \tag{58}$$

is also less than the monoculture yield, however the difference is small. Note that the total amount of nutrient utilized by the end of the season in mixture

$$2\ x(T)/\beta = .98 \tag{59}$$

is greater than for monoculture.

Mixed cropping has often been suggested as a means of increasing yield and decreasing susceptibility to plant disease [Trenbath (1974)]. It is interesting to speculate how the total yield from a mixed crop would change from season to season as the two species adapt from a monoculture to a mixture. Suppose the above mentioned species both have adapted to a monoculture. With $T = 4.25$ it follows that $\beta = 1.45$. Suppose that the number of sites for each species remains constant at

$x_1(0) = x_2(0) = .2057$. Suppose now that in the first season each species grows with its monoculture β. Assume now that from season to season the species adapt one at a time for an increase in yield. Results are given in table 4.

Season	β_1	β_2	$x_1(T)$	$x_2(T)$
1	1.45	1.45	.639	.639
2	1.45	1.35	.602	.652
3	1.25	1.35	.624	.575
4	1.25	1.15	.543	.591
5	1.05	1.15	.553	.507
6	1.05	1.00	.489	.511
7	1.00	1.00	.490	.490

Table 4 Evolution of β.

In the first season, the total yield of 1.278 is greater than the monoculture yield of 1.096. However in the second season species 2 is able to increase its own yield compared to the previous season by decreasing β_2. In the third season species 1 is able to increase its own yield compared to the previous season by decreasing β_1. This process continues until the seventh season when the Nash solution is obtained with its resultant lower total yield.

DEVELOPMENTAL FLEXIBILITY IN MONOCULTURE

By setting $p_2 = 0$ in equation (20) the growth of a single species of a plant under a limited sunlight environment is obtained. For convenience the subscript on p is dropped and a dot is used for a time derivative.

$$\dot{p} = (c_1/(1 + c_5 p) - c_3)p . \tag{60}$$

The total plant matter p can be allocated into a number of subparts, e.g. roots, leaves, seeds, etc. Let u be the % allocation of growth $\dot{p}$ into seed production $\dot{s}$. If we let

$$\ell = p - s . \tag{61}$$

Then

$$\dot{\ell} = (1 - u)\dot{p} \tag{62}$$

$$\dot{s} = u\dot{p} . \tag{63}$$

Assume that the seeds are a small fraction of the total plant biomass. Then (62) may be approximated by

$$\dot{p} = (1 - u)(c_1/(1 + c_5 p) - c_3)p \tag{64}$$

The variable u may now be interpreted in terms of a control variable available via developmental flexibility which may be utilized by the plant to maximize total seed production, i.e.

$$s = \int_0^T u\dot{p}\,dt \,. \tag{65}$$

Cohen (1971) formulated and solved a problem of maximum final seed production for an annual plant based on an exponential model for plant growth. The model given by (64) conforms with one suggested by Cohen (1976) to the author. An extremal strategy for $u(\cdot)$ to minimize (65) subject to (64) under the boundary conditions of a fixed number of "sites" has been obtained elsewhere [Vincent and Pulliam (1977)]. It is sufficient to point out here that the extremal solution for $u(\cdot)$ yields that $u = 0$ at the beginning of the growing season and remains so until reaching the switching time defined by

$$(T - t_s) = (c_1/(1 + c_5 x_s)^2 - c_3)^{-1} \tag{66}$$

where x_s is the biomass at the time of switch. At the time t_s the control switches to $u = 1$ and remains so until the end of the season. Clearly the control program $u(\cdot)$ is a function of the plants current state and time.

DEVELOPMENTAL FLEXIBILITY IN MIXTURE

If two different species of plant are growing in mixture, there is competition for the same sunlight resource as one plant is affected by the presence of the other through shading. In this case the growth of each species of plant is assumed to be given by (20) and (21). Introducing allocation parameters u_1 and u_2 yields

$$\dot{p}_1 = G_1(1 - u_1) \tag{67}$$

$$\dot{p}_2 = G_2(1 - u_2) \tag{68}$$

where

$$G_1 = (c_1/(1 + c_5 p_1 + c_6 p_2) - c_3)p_1 \tag{69}$$

$$G_2 = (c_2/(1 + c_5 p_1 + c_6 p_2) - c_4)p_2 \,. \tag{70}$$

Cooperation is again ruled out by assuming that evolution has provided Nash equilibrium strategies $u_1(\cdot)$ and $u_2(\cdot)$ with respect to seed production, where

$$s_1 = \int_0^T u_1 G_1\,dt \quad \text{and} \quad s_2 = \int_0^T u_2 G_2\,dt \,. \tag{71}$$

Let

$$H^1 = G_1 u_1 + \lambda_3 G_1(1 - u_1) + \lambda_4 G_2(1 - u_2) \tag{72}$$

$$H^2 = G_2 u_2 + \lambda_5 G_1(1 - u_1) + \lambda_6 G_2(1 - u_2) \,. \tag{73}$$

Then necessary conditions for a Nash solution are given by

$$\partial H^1/\partial x_1 = u_1 \partial G_1/\partial x_1 + \lambda_3(1-u_1)\partial G_1/\partial x_1 + \lambda_4(1-u_2)\partial G_2/\partial x_1 = -\dot{\lambda}_3 \quad (74)$$

$$\partial H^1/\partial x_2 = u_1 \partial G_1/\partial x_2 + \lambda_3(1-u_1)\partial G_1/\partial x_2 + \lambda_4(1-u_2)\partial G_2/\partial x_2 = -\dot{\lambda}_4 \quad (75)$$

$$H^1 = k_1 \quad (76)$$

$$u_1 = \begin{cases} +1 & \text{if } \lambda_3 < 1 \\ 0 & \text{if } \lambda_3 > 1 \end{cases} \quad (77)$$

$$\partial H^2/\partial x_1 = u_2 \partial G_1/\partial x_1 + \lambda_5(1-u_1)\partial G_1/\partial x_1 + \lambda_6(1-u_2)\partial G_2/\partial x_1 = -\dot{\lambda}_5 \quad (78)$$

$$\partial H^2/\partial x_2 = u_2 \partial G_2/\partial x_2 + \lambda_5(1-u_1)\partial G_1/\partial x_2 + \lambda_6(1-u_2)\partial G_2/\partial x_2 = -\dot{\lambda}_6 \quad (79)$$

$$H^2 = k_2 \quad (80)$$

$$u_2 = \begin{cases} +1 & \text{if } \lambda_6 < 1 \\ 0 & \text{if } \lambda_6 > 1 . \end{cases} \quad (81)$$

Singular solutions are not obtained in this case. Assuming the initial number of sites fixed for each species results in the boundary conditions

$$x_1(0) = \text{constant} \quad (82)$$

$$x_2(0) = \text{constant} \quad (83)$$

$$\lambda_3(T) = \lambda_5(T) = 0 \quad (84)$$

$$\lambda_4(T) = \lambda_6(T) = 0 . \quad (85)$$

A solution is obtained by guessing values for the final state $x_1(T)$ and $x_2(T)$, integrating (74),(75), (78), and (79) backward in time subject to (77) and (81). At the initial time, the boundary conditions (82) and (83) are to be satisfied.

A comparison between a monoculture and mixture may now be made. Consider a field which has $x_1(0) = .72$ sites for plant 1 and $x_2(0) = .15$ sites for plant 2. Let the growing season be 75 days. Let species 1 be defined by $c_1 = .15$, $c_3 = .02$, $c_5 = .02$ and species 2 be defined by $c_2 = .25$, $c_4 = .02$, $c_6 = .02$. Species 1 in monoculture will switch from $u = 0$ to $u = 1$ on the 37th day producing $s_1 = 96.37$. However if species 1 is grown in mixture with species 2 then a Nash solution predicts that species 1 will switch to seed production on the 21st day and produce $s_1 = 57.48$.

COMMENTS

It has been demonstrated that the process of adaptation can be made an integral part of conventional models. It would seem appropriate to keep this prospect in mind when using biological models for management purposes. The very process of managing a system could introduce effects which can be adapted for.

The calculations have shown that the control strategies associated with adaptation change from a monoculture situation to a mixture situation. This observation may be useful in the prediction of yields for mixed crops over several seasons when only monoculture experimental data is available.

Of particular interest for future study would be to examine if stable Nash solutions exist for the case where the initial conditions are a fraction of the final conditions. Nitche theory [May (1974)] precludes two species utilizing the same resource in exactly the same way.

ACKNOWLEDGEMENT

This work was supported by National Science Foundation Grant No. INT 76-00846; 1976/1977 at the University of Western Australia, Nedlands 6009.

Comments and suggestions on various aspects of this study made by Dr Brian Trenbath (Australian National University), Dr Bean San Goh (University of Western Australia), and Dr Dan Cohan (The Hebrew University) have been most useful.

REFERENCES

Cohen, D., "Maximizing Final Yield when Growth is Limited by Time or by Limiting Resources". *J. Theor. Biol.* 33, 299-307, 1971.

Cohen, D., Communication to the author June 1976.

DeAngelis, D.L., Goldstein, R.A., O'Neill, R.V., "A Model for Trophic Interaction", *Ecology*, Vol.56, No.4, pp 881-892, 1975.

Leitmann, G., *An Introduction to Optimal Control*, McGraw-Hill, New York, 1966.

Leitman, G., *Cooperative and Non-cooperative Many Players Differential Games*, Springer-Verlag, New York, 1974.

May, R.M., *Stability and Complexity in Model Ecosystems*, Princeton, New Jersey, 1974.

Nash, J., "Non-Cooperative Games", *Annals of Mathematics*, Vol. 54, No.2, 1951.

Pianka, E.R., *Evolutionary Ecology*, Harper and Row, New York, 1974.

Pimentel, D., "Population Regulation and Genetic Feedback", *Science*, 159, 1432-7, 1968.

Smith, J.M., *The Theory of Evolution*, Penguin, Middlesex, 1975.

Trenbath, B.R., "Biomass Productivity of Mixtures", *Advances in Agronomy* Vol.26, pp 177-210, 1974.

Trenbath, B.R., Communication to the author, Dec.1976.

Vincent, T.L. and Mason, J.D., "Disconnected optimal Trajectories", *Journal of Optimization Theory and Applications*, Vol.3, No.4, pp 263-281, 1969.

Vincent, T.L. and Pulliam, R.H., "A Controls Approach to Two Problems in Theoretical Biology", to be submitted for publication, 1977.

Lecture Notes in Economics and Mathematical Systems

For information about Vols. 1–95 please contact your bookseller or Springer-Verlag

Vol. 96: O. Moeschlin, Zur Theorie von Neumannscher Wachstumsmodelle. XI, 115 Seiten. 1974

Vol. 97: G. Schmidt, Über die Stabilität des einfachen Bedienungskanals. VII, 147 Seiten. 1974.

Vol. 98: Mathematical Methods in Queueing Theory. Proceedings 1973. Edited by A. B. Clarke. VII, 374 pages. 1974.

Vol. 99: Production Theory. Edited by W. Eichhorn, R. Henn, O. Opitz, and R. W. Shephard. VIII, 386 pages. 1974.

Vol. 100: B. S. Duran and P. L. Odell, Cluster Analysis. A Survey. VI, 137 pages. 1974.

Vol. 101: W. M. Wonham, Linear Multivariable Control. A Geometric Approach. X, 344 pages. 1974.

Vol. 102: Analyse Convexe et Ses Applications. Comptes Rendus, Janvier 1974. Edited by J.-P. Aubin. IV, 244 pages. 1974.

Vol. 103: D. E. Boyce, A. Farhi, R. Weischedel, Optimal Subset Selection. Multiple Regression, Interdependence and Optimal Network Algorithms. XIII, 187 pages. 1974.

Vol. 104: S. Fujino, A Neo-Keynesian Theory of Inflation and Economic Growth. V, 96 pages. 1974.

Vol. 105: Optimal Control Theory and its Applications. Part I. Proceedings 1973. Edited by B. J. Kirby. VI, 425 pages. 1974.

Vol. 106: Optimal Control Theory and its Applications. Part II. Proceedings 1973. Edited by B. J. Kirby. VI, 403 pages. 1974.

Vol. 107: Control Theory, Numerical Methods and Computer Systems Modeling. International Symposium, Rocquencourt, June 17–21, 1974. Edited by A. Bensoussan and J. L. Lions. VIII, 757 pages. 1975.

Vol. 108: F. Bauer et al., Supercritical Wing Sections II. A Handbook. V, 296 pages. 1975.

Vol. 109: R. von Randow, Introduction to the Theory of Matroids. IX, 102 pages. 1975.

Vol. 110: C. Striebel, Optimal Control of Discrete Time Stochastic Systems. III. 208 pages. 1975.

Vol. 111: Variable Structure Systems with Application to Economics and Biology. Proceedings 1974. Edited by A. Ruberti and R. R. Mohler. VI, 321 pages. 1975.

Vol. 112: J. Wilhlem, Objectives and Multi-Objective Decision Making Under Uncertainty. IV, 111 pages. 1975.

Vol. 113: G. A. Aschinger, Stabilitätsaussagen über Klassen von Matrizen mit verschwindenden Zeilensummen. V, 102 Seiten. 1975.

Vol. 114: G. Uebe, Produktionstheorie. XVII, 301 Seiten. 1976.

Vol. 115: Anderson et al., Foundations of System Theory: Finitary and Infinitary Conditions. VII, 93 pages. 1976

Vol. 116: K. Miyazawa, Input-Output Analysis and the Structure of Income Distribution. IX, 135 pages. 1976.

Vol. 117: Optimization and Operations Research. Proceedings 1975. Edited by W. Oettli and K. Ritter. IV, 316 pages. 1976.

Vol. 118: Traffic Equilibrium Methods, Proceedings 1974. Edited by M. A. Florian. XXIII, 432 pages. 1976.

Vol. 119: Inflation in Small Countries. Proceedings 1974. Edited by H. Frisch. VI, 356 pages. 1976.

Vol. 120: G. Hasenkamp, Specification and Estimation of Multiple-Output Production Functions. VII, 151 pages. 1976.

Vol. 121: J. W. Cohen, On Regenerative Processes in Queueing Theory. IX, 93 pages. 1976.

Vol. 122: M. S. Bazaraa, and C. M. Shetty, Foundations of Optimization VI. 193 pages. 1976

Vol. 123: Multiple Criteria Decision Making. Kyoto 1975. Edited by M. Zeleny. XXVII, 345 pages. 1976.

Vol. 124: M. J. Todd. The Computation of Fixed Points and Applications. VII, 129 pages. 1976.

Vol. 125: Karl C. Mosler. Optimale Transportnetze. Zur Bestimmung ihres kostengünstigsten Standorts bei gegebener Nachfrage. VI, 142 Seiten. 1976.

Vol. 126: Energy, Regional Science and Public Policy. Energy and Environment I. Proceedings 1975. Edited by M. Chatterji and P. Van Rompuy. VIII, 316 pages. 1976.

Vol. 127: Environment, Regional Science and Interregional Modeling. Energy and Environment II. Proceedings 1975. Edited by M. Chatterji and P. Van Rompuy. IX, 211 pages. 1976.

Vol. 128: Integer Programming and Related Areas. A Classified Bibliography. Edited by C. Kastning. XII, 495 pages. 1976.

Vol. 129: H.-J. Lüthi, Komplementaritäts- und Fixpunktalgorithmen in der mathematischen Programmierung. Spieltheorie und Ökonomie. VII, 145 Seiten. 1976.

Vol. 130: Multiple Criteria Decision Making, Jouy-en-Josas, France. Proceedings 1975. Edited by H. Thiriez and S. Zionts. VI, 409 pages. 1976.

Vol. 131: Mathematical Systems Theory. Proceedings 1975. Edited by G. Marchesini and S. K. Mitter. X, 408 pages. 1976.

Vol. 132: U. H. Funke, Mathematical Models in Marketing. A Collection of Abstracts. XX, 514 pages. 1976.

Vol. 133: Warsaw Fall Seminars in Mathematical Economics 1975. Edited by M. W. Loś, J. Loś, and A. Wieczorek. V. 159 pages. 1976.

Vol. 134: Computing Methods in Applied Sciences and Engineering. Proceedings 1975. VIII, 390 pages. 1976.

Vol. 135: H. Haga, A Disequilibrium – Equilibrium Model with Money and Bonds. A Keynesian – Walrasian Synthesis. VI, 119 pages. 1976.

Vol. 136: E. Kofler und G. Menges, Entscheidungen bei unvollständiger Information. XII, 357 Seiten 1976.

Vol. 137: R. Wets, Grundlagen Konvexer Optimierung. VI, 146 Seiten. 1976.

Vol. 138: K. Okuguchi, Expectations and Stability in Oligopoly Models. VI, 103 pages. 1976.

Vol. 139: Production Theory and Its Applications. Proceedings. Edited by H. Albach and G. Bergendahl. VIII, 193 pages. 1977.

Vol. 140: W. Eichhorn and J. Voeller, Theory of the Price Index. Fisher's Test Approach and Generalizations. VII, 95 pages. 1976.

Vol. 141: Mathematical Economics and Game Theory. Essays in Honor of Oskar Morgenstern. Edited by R. Henn and O. Moeschlin. XIV, 703 pages. 1977.

Vol. 142: J. S. Lane, On Optimal Population Paths. V, 123 pages. 1977.

Vol. 143: B. Näslund, An Analysis of Economic Size Distributions. XV, 100 pages. 1977.

Vol. 144: Convex Analysis and Its Applications. Proceedings 1976. Edited by A. Auslender. VI, 219 pages. 1977.

Vol. 145: J. Rosenmüller, Extreme Games and Their Solutions. IV, 126 pages. 1977

GPSR Compliance

The European Union's (EU) General Product Safety Regulation (GPSR) is a set of rules that requires consumer products to be safe and our obligations to ensure this.

If you have any concerns about our products, you can contact us on ProductSafety@springernature.com

In case Publisher is established outside the EU, the EU authorized representative is:

Springer Nature Customer Service Center GmbH
Europaplatz 3
69115 Heidelberg, Germany

Batch number: 10704763

Printed by Printforce, the Netherlands